'A wonderful, beguiling book, inviting us back home to nature, from which only delusion has estranged us.'

Robyn Davidson, author of *Tracks*

'It reads like a love letter to nature. I felt I was on a journey into becoming a listener, and I didn't want the journey to stop.'

Dr Sue Gould, conservation ecologist

'I was very moved reading this, I feel Andrew offers people a whole new insight into our relationship with the world. This book will be loved by many, and help us return closer to the Earth.'

Brian Walters, AM SC

'This fascinating book is a convincing argument for why we must listen to our environment and all within it. It also shows us the pleasure and privilege which awaits anyone who is willing to listen deeply.'

Dr Lynne Kelly, AM, author of
The Memory Code, Memory Craft

'I loved *Deep Listening to Nature* and found it both meditative and moving to read – an antidote to the trauma of world events going on around us. The vicarious travel to some of the most remote areas of the world was a wonderful added bonus – a peek not just at the landscape, but the biodiversity and sounds of the natural habitat. What a joy.'

Kristin Gill, publishing consultant

Together with his partner, photographer Sarah Koschak, Andrew established the independent label Listening Earth in 1993 to publish immersive nature soundscape recordings. This work has since taken him around the world, documenting the sounds of iconic landscapes and threatened ecosystems.

The resulting recordings have been published as likely the largest catalogue of their kind; over 100 nature albums available for online download and via digital music platforms such as Spotify, resulting in streaming figures in the many tens of thousands a week.

Andrew's personal yet broad perspective has led to invitations to talk to a wide range of audiences. He was recorded by ABC radio for its *Big Ideas* program, and in 2017 presented at TEDx in Canberra. He has been a keynote speaker at academic conferences, and regularly teaches at university, schools and for community organisations.

Andrew is president of the Australian Wildlife Sound Recording Group, a premier association of nature field recordists that encourages skills and passion in a new generation of enthusiasts.

Deep
Listening
to Nature

ANDREW SKEOCH

Listening
Earth

Published in Australia by Listening Earth,
PO Box 144, Newstead, Victoria 3462, Australia
listeningearth.com

First published 2023

This book has been written in good faith.
The publisher apologises for any inaccuracies, omissions
or lack of attribution, and would be grateful if notified of
any corrections that could be included in future editions
of this book.
andrewskeoch.com

A catalogue record for this
book is available from the
National Library of Australia

NATIONAL
LIBRARY
OF AUSTRALIA

ISBN 978 0 6457563 0 2 (pbk)
ISBN 978 0 6457563 1 9 (ebk)

Cover image, 'Southern Scrub-robin', © Lachlan Read,
Instagram.com/lachlan.read
Typeset by Helen Christie, Blue Wren Books
Printed by Ingram Spark

This book has been written where I now live in central Victoria, in southeast Australia. These are the traditional lands of the Dja Dja Wurrung peoples of the Kulin Nation. I acknowledge their elders and wisdom keepers, and their custodianship of the land over so many generations.

I have been privileged to travel widely, and am aware that many places I've been in the world have similarly been traditional land for First Nations Peoples of one culture or another. Some have long been subsumed into colonising cultures, others continue in the modern world. I wish to thank those from whom we've received hospitality, and acknowledge all who strive to care for natural places and the wild creatures who inhabit them.

For Sarah

who has shared this journey with me

Contents

Get your Headphones Ready

This book relates my experiences of listening to the natural world. To bring these stories alive and as evidence for the resulting ideas, I need to let you hear what I've heard.

Hence the text is accompanied by audio files, available online. These recordings are in stereo, many of them binaural and conveying a rich sense of space, so headphones will be the optimal way to enjoy them.

Using the link or QR code below will take you to a webpage, with chapter headings. Clicking chapter links will reveal the audio materials relevant for each chapter. Audio files are sequenced as they occur in the text, and as you read, references to relevant recordings are indicated by the symbol: 🐦

You may like to open the webpage on a phone, tablet or laptop and listen as you go, or 'bulk' listen before or after reading a chapter.

However it works for you, listening to these recordings is an important aspect of this book, and will give you the auditory experience that words cannot.

https://listeningearth.com/deeplistening/

Prologue

The aromas of native pine blossom, emu bush and humid earth scent the night air, as I set off on foot in the darkness before dawn to climb a low ridge in the Australian outback. All around, the bush is silent, still, expectant. Unseen, and secretly tucked into shrubs and tangles of foliage, birds are roosting. Are they asleep, I wonder, their bills nuzzled into warm feathers against the cold? Or are they furtively awake, eyes bright and watching the stars above for the first hints of approaching daylight, their signal to begin singing and usher in a new day?

As I begin finding my way up the slope, I reflect on what I hope to achieve here. Sarah[1] and I have come to this place, Mutawintji, in the far west of New South Wales, to pursue a music recording project. In our few years together, we've connected over a love of camping and spending time out in the bush, during which we've discussed ways of shaping our interests into a creative vocation. To this end, we've established a fledgling record label, Listening Earth, with the intention of releasing music inspired by a connection with nature. As a musician, I've been composing classically styled music for acoustic instruments, and our project envisages these pieces being heard in the context of the natural environment. Mutawintji has appealed to us as a suitably inspiring location in which to pursue this idea. So now, after vaguely dreaming of it previously, I am beginning to make my first birdsong and nature sound recordings.

Even in the few days we've been here though, things have not gone to script. Mutawintji is certainly turning out to be a special area; a low range of hills outcropping from the surrounding sand-plains, its secluded gorges holding permanent water in rockpools, providing a haven for wildlife in otherwise arid country. However, the desert is living up to its reputation for extremes. The night after we arrive, a storm front sweeps in, and while making my very first recording, I nearly fry the electrical gear (and myself) when a lightning discharge splits the air nearby. Lesson one for the novice nature recordist; don't monitor on headphones when there's thunder around!

That night, the heavens open and chuck it down. The following morning dawns clear, but my recordings of the usually dry Australian outback feature torrents of water flushing down gorges and stream beds, to flood out across the open plains. While this is sonically interesting, it is not an ideal situation, as the wildlife is following the water. No longer conveniently concentrated in the sanctuary of the gorges, the birds are dispersing out across the landscape. Now I'm guessing the ridge top will be as likely a place as any to try recording the dawn birdsong.

To be honest, I have little idea what I might hear, as I'm still familiarising myself with the local environment. I'm also working out how to make nature recordings with the gear we have. Here at least, serendipity has smiled in the form of a friend able to lend us a complete rig of field recording equipment. This comprises a new generation, portable, digital audio recorder – a vast improvement on the heavy, analogue, reel-to-reel equipment previously available – plus a pair of long, 'shotgun' type microphones. With a highly directional pickup, I am already realising they aren't particularly suitable for making balanced stereo recordings, but for the moment, they are all we have. All this is safely stored in my backpack, along with cables, spare batteries, tapes and a pair of headphones.

I pause, turning to look out across the landscape stretching to the west and dimly visible by starlight. Apart from the ranger station and guest quarters where we're based, and from which I'd set out earlier, there is no human presence to be seen. The open desert stretches away to the horizon. It is palpably quiet.

Arriving at the ridge top to the first flush of pink in the east, I set up the gear. The ground is rocky and open with a few hardy acacias, and the air is crisp and still. From somewhere out across the plains, a magpie is now faintly audible, warbling a way off. The mobs of corellas which roost in the river gums lining the gorges are also waking, their screeches softened by distance. Perfect timing – the dawn chorus is just beginning.

After listening across the expanse of landscape to these far away voices for ten minutes or so, a flutter of wings nearby surprises me as a bird flits into a shrub. Almost immediately it utters the most entrancing sound; a quick series of piping, flute-like notes, ever so slightly descending in pitch. The song lasts only a few seconds and is hauntingly beautiful. I've heard nothing like it so far and have no idea what species it is. Presently the bird sings again, its voice a warmly-toned cadence in the desert air. Soon, another begins responding from a little way along the ridge, their voices sometimes alternating, at others entwining in harmonious microtones. Further off, I can now discern others; there is quite a dispersed community of these birds singing together across the ridges.

The combined effect of their songs is ethereal. As I continue listening, a sense comes over me of them being so native to this area that they are giving voice to the landscape itself. By belonging to these rocky ridges and gorges, their songs are animating the country, singing it alive. The distant magpies and corellas also contribute, their voices combining to uniquely signify this place. It occurs to me that, as this is a daily happening, I'm eavesdropping on something ancient, essential, timeless. Hearing not only the beauty of my mystery species, but this blending of birdsong born of the land, I feel unexpectedly moved.

Eventually, another flutter of wings signals my sublime flautist moving off. Then, with a last few calls from further away, I sense the dawn chorus subsiding. Shortly after, I switch off, and sit down for a long time as the light grows, letting what I've heard replay in my memory.

I wonder why hearing the beauty of the birdsong here comes as such a surprise. I've spent so much of my time in the bush. Even though I grew up in suburban Sydney, nature has always been a fascination, and I've been actively birdwatching throughout my childhood. But of course, this is it; I've been watching. Often using binoculars, I've been referencing field guidebooks to identify species by form and plumage. My knowledge of birds has been entirely visual, not auditory. With my interest in music, I would have thought I was proficient at listening, and perhaps I am – after all, I wouldn't be here recording if I wasn't tuned in to sound already. Yet, I am deeply affected by what I have heard this morning, feeling that an essence of the natural world has been waiting, as obvious as anything, until this moment when I am ready to hear it.

Back at our quarters, I catch up with Sarah. Already, we are falling into roles; myself sound recording while she, the more visual of us, has been out photographing. In letting her hear my recording, I realise that our initial idea of gathering general bird-song ambiences is now gaining the substance of actual species and their characteristic voices. Over the ensuing days and weeks at Mutawintji, I immerse myself in listening, increasingly captivated by what I am hearing and recording.

During this time, we get to know the ranger, Sharon, who lives in the adjacent homestead. She is a sunny spirit with a deep love of the place. The only other resident is Harold, an older Aboriginal man employed doing general tasks around the park. He shares the shearer's quarters with us, and while always friendly, seems self-contained and mostly keeps to himself.

Mutawintji, apart from being an outstanding natural area, is also a site of cultural significance for the local Barkindji Aboriginal people. Before colonisation, for generation upon generation, regional groups had gathered here to feast, share culture and socialise. While their contemporary descendants live mainly in nearby towns, the landscape remains alive with the presence of those ancient peoples, most obviously in the form of numerous, well-preserved rock art and petroglyph sites.

Sharon tells us that Harold knows the park like no one else, walking alone for days to outlying areas in search of forgotten art sites and ceremonial locations. We get the sense that he is a deep spirit, that Harold. He's also evidently curious about what we are doing, setting off as we do at odd hours and often being out all day.

One afternoon, Harold comes over and asks if we'd like to accompany him, telling us he's going down to the southern end of the range. As we've not had many previous opportunities to hang out with Aboriginal people, we welcome the opportunity.

Setting off together, we drive the dirt road that runs parallel to the range. Harold says little. It doesn't seem the occasion for small talk. For half an hour, we rattle along the corrugated track, which eventually takes us closer to the hills and terminates at the entrance to an impressive gorge. The engine is switched off, and alighting, we stand in the stillness, taking in our surroundings. I am already listening, hearing a Willie Wagtail calling from somewhere up on the stony slope.

Harold turns to us. "We're going to find a place to sit quiet. You go maybe on those rocks over there," he says, indicating with a gesture of his hand. "We'll just sit awhile and listen."

After the drive, I am glad of the opportunity to settle, get the sounds of the vehicle out of my head, and let my ears adjust. I nod, "Good, it'll give us some quiet to tune into the bush."

He focuses on me for a long moment, as though wondering how or even whether to respond, then says softly, "No, that's not why we do this."

Now I am puzzled. "I don't understand."

"This is not for you to tune into the bush," he continues. "It's for the bush to tune in to you. Find out what kind of fella you are. Whether you're of good character, whether you can be trusted. If you are, the bush will start revealing itself. Start talking to you. It'll tell you things."

These experiences occurred many decades ago, when I was in my early thirties. They marked the beginning of what would turn out to be a life of listening to nature. As Sarah and I sat quietly that afternoon, with Harold a short distance away, I thought of his Aboriginal perspective as a lyrically poetic yet mildly superstitious way of perceiving nature. I didn't appreciate then the depth of his words, and just how profoundly connected was his relationship with the country around us. Nor did I recognise the habits and assumptions that pervasively disconnect my own culture.

In time, I'd come to reflect on Harold's words as embodying an essential wisdom, one pointing the way to appreciating the communicative relationships that infuse the natural world, and to finding my own place among them.

Chapter 1

An Invitation to Listening

Deeply listening to another person, particularly one whose life experiences, beliefs or world view differs from our own, has been described as a radical act. As a concept in Western thinking, it emerged in the 1980s from the contemplative practice of artists such as Pauline Oliveros[2] and Hildegard Westerkamp.[3] More recently, the skills of deep listening have been applied to social engagement, personal relationships and group counselling.

If deeply listening to another person can be considered challenging, then the listening journey I propose is equally, if not more so, as it will take us beyond the human world altogether. Beyond the values, ideas, opinions, news, information and entertainments that fill our days as highly social animals. Beyond the daily conversations that we share with each other. Instead of inward, to ourselves, to the familiar, I shall be encouraging you to listen outward; beyond our species, our human bubble, outside of what is sometimes referred to as the anthroposphere.

My invitation is to listen deeply to the natural world, to those multitude of other beings with whom we share our planet.

Perhaps this may seem a straightforward exercise, after all, hearing is such a fundamental sense of perception that we use every day.[4] Yet its very familiarity can cause us to underestimate it. The act of truly listening involves far more than simply hearing. Deep listening requires not only *attention* to what we hear, but the *intention* to open our senses, and allow what we perceive to influence us. By doing so, we learn as much about ourselves as the other. This can be confronting enough and generate profound outcomes when we listen deeply to another person. If we listen to nature with a similar intention, the context is broadened to *all* of life. To find meaning in the communications of other creatures with whom we don't share a commonality of kind, requires not only learning the languages of nature, but cultivating a more attentive mode of listening.

I can say this clearly now, but I was far from aware of it initially. As Sarah and I hiked and explored under a wide Mutawintji sky, I was not thinking about the act of listening. Instead, always ready with the microphones, I was opportunistically seeking to capture anything interesting. Playing back those recordings now, I can remember the excitement and fascination I had when making them. I can bring to mind the person I was, discovering this new 'window' on nature, and with the naive enthusiasm of youth, envisaging how they would contribute to our creative project. Now, with headphones on and eyes closed, I can even recall the moment; my surroundings, the light, the ground underfoot, the dry desert air. This is the poignancy with which sound can bring memory alive.

However, with some dismay, I'm also aware that many of the recordings I made then were short, often a few minutes here, another few minutes there. The longest recording – yes, on that ridgetop – was only twenty minutes. I was capturing sound bites. At the time, I imagined assembling the 'good bits' into a sonic portrait of the environment, like a set of acoustic postcards. This approach seemed thoughtfully conceived and self-evident to me then. Now I recognise I was carrying the presumptions learned from the culture in which I'd grown up. My perception of nature's soundworld was fragmented because my listening was similarly inconsistent and unsustained.

How can we cultivate a deeper sense of listening to the natural world? Fortunately, we're wired to do so. The human brain itself is a product of nature. By default, our cognition is tuned to the sounds of the living world. Throughout a long, long evolution over countless generations of our animal predecessors, the minds that eventually became human have been immersed in the sounds around them. Nature has provided the sonic stimulus to which our sensory perception has refined itself.

For humans, like any animal, the neural connections required in listening are reinforced through daily life. Our distant ancestors, living in close connection to nature, would likely have been aware of their surroundings with far greater acuity than we practice today. As they walked the land, they would have used sound to assess whether there were opportunities, or threats, to respond to. One expression of this is that while we may expect our hearing to be adapted to the voice, my colleague, the American sound recordist Lang Elliott,[5] has suggested that instead, the human ear is acutely sensitive to the frequencies and nuance of rustling grass. From my experiences of wildlife moving through the dry, waist-high grasses of East Africa's savannahs, the habitat in which pre-human cognition evolved, I can fully appreciate why that would be.

Once language began emerging in early *Homo sapiens*, listening would have remained vital as increasingly complex communications within our species developed. In societies without writing, practicing what is termed primary orality, all knowledge had to be heard, memorised and passed on precisely. Learning came from elders telling stories, singing songs and describing the interactions of the living world. In an oral culture, accurate listening, comprehension and recollection skills would have been life skills. They would have ensured survival.[6]

You'd have to assume those peoples were far more keenly aware of the sounds around them than ourselves as modern listeners. This is suggested by the way Indigenous cultures often recognise and name creatures by their voices rather than appearance. Australian Zebra Finches for instance, named by European colonists after an obscure visual reference to a mammal not even native to the continent, are known by Western Desert Aborigines as 'nyi-nyi's, a fair approximation of their call. 🐦 The Indigenous name 'currawong' has been applied by science to the entire *Strepera* genus, yet only one of its member species, the Pied Currawong, gives that evocative cry. 🐦 Across the country, crows and ravens have many names in Indigenous languages, but variations of 'Kwaa', 'Karrnka' and 'Waa' are common.

These close associations with everyday sounds are an indication that, for subsistence hunting and gathering peoples, listening to their environment would have been a life or death skill – not only necessary, but unquestioned.

When one's listening is this attentive, nature is experienced as an intimate reality. In this kind of intimacy, listening becomes a transcendent practice. It becomes more than simply the hearing and objective interpretation of sounds. Rather, it opens up a communion. We become present to the world through the act of listening. We simultaneously hear the natural world in the moment, and are aware of our relationship to it. We become part of what is heard.

Slowing Down to Listen

All this suggests that, when it comes to listening to nature, our twenty-first century cognition is poorly adapted. The skills that accommodate us to a technological world of rapid information delivery do not serve us well when it comes to being aware of the natural world.

Essentially, the issue is this: nature operates at a different pace.

In nature, sonic processes often happen on a gradual and diffuse time scale. In the natural soundscape – the sum of all sounds that can be heard in any one place and time[7] – events such as a dawn chorus may emerge, grow, sustain and then dissolve over several hours. Other patterns manifest on even broader timescales, being governed by daily or seasonal cycles. Even when listening to the moment by moment flow of what is audible, change may happen imperceptibly. One may not notice immediately that a new voice has emerged in the landscape, or that a prevalent one has fallen silent.

To listen to nature mindfully, we have to slow down. We must adapt to an unhurried tempo shaped by a schedule not of our own making. We need to switch down the gears. While we may be used to streaking along in a cognitive fast lane, weaving and overtaking, nature requires something more pedestrian of us. Not only slower, but attuned to nuance, to the delicate, the subtle and the ephemeral.

This refocusing of our attention can be elusive. I certainly didn't find it easy when I first began, habituated as I was at the time to urban life. Of course, cities are pervasively filled with low-significance sound, and our public spaces often cluttered with distracting acoustic stimuli. Filtering all this input for what is relevant to us and what is not, minute by minute and day after day, puts considerable cognitive demands on the brain. Our minds achieve this acoustic processing very efficiently, yet many will readily identify a connection between noise and stress.

Hence it is understandable that in urban circumstances, taking the time to actively listen to one's surroundings can become rare. For many of us, it may constitute a brief and occasional moment of curiosity – to a bird singing in the garden perhaps, or distant frogs that have begun calling after rain – a slight noticing of the world around us.

This desensitising – literally 'de-sensing' of ourselves – to the natural world seems a symptom of contemporary life. It affects us all, as my own experience revealed to me. When nature assumes only a peripheral importance, we lose those cognitive functions that allow us to comprehend it. They atrophy. Ironically, this is a natural plasticity, the capacity of our brains to reshape themselves according to need – for us, the forging of neural pathways in response to our technological lives.

Yet nature is the sonic world we have evolved to make sense of. The capacity to attune to the pace of nature and the patterns of organic sound, remain deeply embedded in our biological minds. As we nurture a fresh focus on listening, we can be confident that the mind's inherent flexibility will allow latent capacities to be re-awakened, gently orienting us to perceiving our natural surroundings anew.

The Sense of Hope

The way we listen is thus an expression of the way we live.

I find it no coincidence then, that contemporary Indigenous cultures, with their traditions of listening to the land, find themselves at the forefront of advocating for environmental protections. In the same way that radical listening implies respect for the person being heard, deep listening to nature also embodies respect. Honouring is an inescapable consequence of being aware. These two themes, foundational to the thinking of First Nations Peoples, seem to echo down the millennia; listen to the land, care for the land.

In this way, listening offers hope.

Firstly, it does so personally. By fostering our listening skills, we can enrich our connection with nature. This can be a lot of fun and engage us at any age. I recently had the opportunity of working with a group of primary-age school kids on a Melbourne bayside excursion. They were a good natured but typically unruly mob, yelling and play jostling one another. Out on a pier, with hydro-phones (special microphones for listening underwater) dangling down into the depths below, they listened on headphones with complete attention. Immersed in a new soundworld, they became children rapt in fascination, listening to the sharp clicks of marine shrimp for the first time. 🐦 Their faces told the story. Listening offers an immediate, sensory and emotional connection, engaging our being with the living environment.

Secondly, listening offers humankind hope of renewal. As our oldest cultures teach, taking the time to listen is a way of caring. Implicitly, we are showing respect for the natural world by giving it our attention. Through this intention to listen, we are not passive, but actively creating something – a relationship between ourselves and what we become aware of.

Conceiving of listening in this manner quickly runs smack bang into entrenched assumptions. In the modern, industrial world view, with its insistence on objectivity, any implied relationship with nature is suspect. Surely, when we hear wind in the trees it is simply the audible consequence of an inanimate process? Birdsong may represent communication, but aren't birds themselves largely indifferent to our listening presence? Where is the relationship?

With this scepticism, we maintain objectivity, and with it, disconnection. True listening – deep listening – is incompatible with separation. It subverts withdrawal by returning us to our senses, immersing us like children in the breathing, singing, murmuring, whispering, animate world. Through listening, we can transcend the self-defined boundaries of our anthroposphere and enter into the communicative life of the biosphere.

I suggest that intimate listening is thus the most completely human way of relating to the wild and free-living creatures with whom we share our planet. Maybe in doing so, we can unlearn those habits of our self-referencing culture, and find a new sense of our human place in the world.

If we're to fully embrace what deep listening has to offer us, we will need to be open to learning as much about ourselves as nature itself.

Our Listening Journey

In one way or another, this book addresses a question I've been asking most of my life: what is our true relationship to nature? – both as individuals and as a species. In coming to my own answers, I've found that listening, by providing a direct awareness of nature, brings us back to the fundamentals of relationship, and thus offers a unique and fresh perspective.

I need to begin this book by assuming that my invitation to listen will come as a novel adventure. So we'll start by easing into a sound-focused awareness; tuning our ears in to nature and the daily lives of creatures around us. I'll discuss the field skills of how to identify species by ear, interpret their sonic behaviours and appreciate their sentience.

From there, we can explore a central issue; what listening reveals about how the living system of nature functions. In pursuing this, I'll be taking you on a global journey to immerse ourselves in the mysteries of deep time, and explore how creatures have evolved the use of sound and communication to regulate their interactions. In particular, the songs of birds will allow us to consider how they negotiate their relationships in such lyrical ways, and the purposes they achieve by doing so.

Interspersed with the above topics will be short interludes, in which I'll share pivotal moments in my own listening journey.

All this will form a necessary foundation for the concluding section of the book, in which I'll reflect on what we may learn from nature in the context of our current environmental challenges. How may we mimic what nature has achieved in sustaining life, as we move toward an ecological future?

The Limitation of Words

Before we launch into things, there is a final issue I need to mention. It is one with breathtaking implications for the way we understand the world, and it's this: books don't do sound.

When you consider that, since the advent of writing, our collective human knowledge has been conveyed predominantly through text on the page, and that those pages have been incapable of communicating the acoustic realm, we get an inkling of just how limited our understanding of the world has become. The page simply can't do justice to sound in the way that it can to the visual, where images, drawings, maps, graphs and diagrams can be included. When we convey information, one of our most crucial sources of meaning can only be poorly and obliquely referred to.

This failure of the medium is amplified by the paucity of language we have to describe the subtleties of abstract sound. In speech, meaning is conveyed through inflection, intonation and dynamics, and we have a reasonable language for conveying these nuances in prose. However with natural sounds such as birdsong, where subtle distinctions can differentiate species, subspecies or local populations, we do not have a comparable wealth of vocabulary. Trying to articulate the character of a particular creature's voice can be fraught with vagueness. We can employ words such as metallic, liquid, grating, silvery, harsh, ringing and so on, but these somewhat poetic allusions are subjective, and may only be useful if you appreciate the sonic quality being described.

The written word also acts to diminish our appreciation of the life force of our fellow creatures. When a creature vocalises, we hear it giving voice to itself. As animals, we listen, and there is an empathic response in hearing another living being. We can perceive something of its life essence, the spirit of the animal in its voice. If we have insufficient language to describe sound, we have even less to articulate this essential connection that listening affords us with other creatures.

Words thus fail us. There are times though, when we are reminded of how primal sound can be. In Tanzania, Sarah and I observed at close quarters from the safety of our vehicle as a pride of Lions tore apart the carcass of a recently downed Wildebeest. It was a gruesome spectacle, yet it was not the sight that was affecting. It was the continual sound of panting and low grunting, of bones being crushed and sinews torn. 🐦 Several hours later, with the cubs finishing off the last scraps, the pride of adults began roaring in chorus. They were by now lying all around us on the ground, and with necks outstretched and bodies taut, they heaved out each breath, creating a resonant, guttural sound. 🐦 After hours of watching in fascination, their roars were visceral. My knees went involuntarily weak and I suddenly felt acutely vulnerable. It was an instinctive, gut reaction.

This is the power of sound. Its physicality alone conveys meaning.

These disparities between text and sound are due to our senses functioning differently. In seeing, we divide the world up, focusing our vision on discrete objects and recognising them as separate. These visual objects initiate language; we name things. When language hits the page, ideas become, somewhat literally, set in stone. This subtly habituates us to objectifying the world and perceiving it as static.

Sound, on the other hand, is dynamic, an expression of the ever-changing flow of life. When listening, one can analyse in terms of 'sound objects', yet, as with the notes of music or syllables of words, it is their relationships and interactions that convey meaning. In addition, our hearing is not limited by a specific direction nor focused at a particular distance. It is spatially holistic, our ears taking in everything at once; near and far, loud and soft, in all directions.

So whereas the visual, linguistic and written generate inevitable separations, when listening, we put the world back together again.

Chapter 2

A Practice of Listening

For many, I suspect the mention of 'nature' will conjure up a place. It may be a neighbourhood park, a hiking trail through a wild landscape, or a shady tree under which to stretch out for an afternoon snooze. It may be an ocean beach or favourite fishing spot. Often it will be a familiar and comforting place, possibly one remembered from childhood. Even when we recognise habitats or environments as home to wildlife, we are still perceiving nature, on some level, as a place.

Perhaps this is a conceptual habit of Western culture, in which land and its ownership mean so much. But whatever the reason, it is a subconscious association I need to point out, because throughout this book, the 'nature' I shall be inviting you into is vibrant with the relationships and interactions of living things. Rather than a noun, nature is a verb. Rather than a place, it is a process.

Listening is an especially appropriate means of perceiving nature in this way, because sound results from an action or event of some kind, and so tells us what is happening. This gives us a fundamentally different perspective – one that speaks of activity, dynamics, relationship and interconnection.

Listening gives us an animate awareness of the natural world.

A Listening Awareness

This auditory awareness of nature needs to be cultivated however, because while the ears hear, it is the mind that listens. Our brain itself is a complex expression of nature, and cognition functions in complimentary modalities. Hence there are multiple skills to be nurtured, which can be simplified as two; one perceptual and the other interpretive. We must first train our listening awareness, and then find meaning in what we hear.

Developing our perceptual skill is perhaps the more challenging, as it's fundamentally about being present and noticing, so I'll address it first. To begin focusing this listening mind, I offer the following suggestions. For me, they've become habitual, and in articulating them, I'll try and convey what I do without really thinking about it. None are obligatory, nor what I'd suggest you do every time – they're not a regime. Rather, they're perspectives on listening, aspects to explore and reflect on, ideally until they become familiar and integrated – second nature you might say.

At this point I'd like to invite you outdoors to listen. The kind of location I'm imagining is one in which nature, in whatever degree of wildness, dominates over human presence. Listening already gives us a measure of this – you'll likely recognise it as a 'quiet' place.

It will also be a place that is alive. And this is where we should always begin, by reminding ourselves of what is essential – that we too are alive. We are living beings, inhabitants of the biosphere, and nature is our place of belonging. As you become aware of what's around you, consider that every organism also has its own ways of perceiving its surroundings. As you sense, you too are being seen, heard and regarded. So enter gently into any wild environment with an openness. As was suggested to me – allow time for nature to get to know you.

A listening awareness is part of our natural inheritance. However as modern peoples, we're often habituated to a kind of specific, single-pointed listening, such as when focused on a conversation while filtering out much of our ambient surroundings. Nature requires that we develop a more encompassing mode of listening, easing past our perceptual filters and awakening ourselves to everything around us. I'd describe it as extending my sense of hearing into the surrounding space – spatial listening, if you like. This may come as something quite unfamiliar, and initially require some conscious intention.

So begin by being physically still and quiet, taking in the sound-scape around you. As you do, notice any sounds coming from left or right, in front or behind. Without moving physically, scan your awareness around in a complete circle, taking in the listening space all about. Also sense the hemisphere above; the air, trees and sky overhead. It's possible there may even be sounds below; closer, intimate sounds from the ground near you. Note the directions from which sounds are emanating, and conversely, the places, possibly whole quadrants, that seem quiet. Listen into those empty spots, sensing your way into their quietness, possibly even finding they're not entirely silent after all. This encourages the extended listening I'm suggesting. In expanding your perception into the whole acoustic space, you'll feel yourself becoming present in the landscape, centred in a physical locale.

Now take account of distance. How far can you hear? What is the furthest sound audible? You're now defining the acoustic horizon, your circumference of perception. Atmospheric conditions will affect this horizon greatly. In still conditions and with lower temperatures, such as early mornings, sound may travel from surprisingly far away, but by midday, it will be significantly contracted. Ambient sound, such as wind in trees, flowing water or beach surf, may mask and reduce the acoustic horizon markedly. Topography or prevailing wind direction will often lead to an asymmetric horizon, extending further in some directions than others.

You may at first think that you're hearing sounds 'in the landscape', but landscapes themselves have their own acoustic. This is the reverberation, or echo, produced by vegetation and landforms. Each ecosystem has a unique sonic character. Forests and woodlands generate a smooth, even reverberance, as sounds are diffused and reflected through vegetation. Harder echoes, perhaps with a noticeable time delay, may be created by reflections off rockwalls or tree lines. Conversely more open landscapes such as grasslands, coastal areas or deserts will often have a very 'dry' sense of space, giving the impression of sounds travelling across the landscape, crisp and altered only by distance.

As a field recordist, I love capturing the sonic properties of the landscape with my microphones, as they enhance a recording's ability to convey a picture for the listener. They also have significance biologically. For any species, the acoustics of the habitat in which they live will be integral to how they communicate. Some go further, and utilise the acoustic of their surroundings in specific ways. For instance, Canyon Wrens in the Southwestern deserts of America use cliff walls and rock overhangs to amplify their voices, and no hearing of them would be authentic without that canyon reverberance. 🐦 So while it may be elusive to hear this landscape acoustic distinctly, see if you can sense the flavour of it.

Along with reverberation, distance also alters sound. By the time you hear them, voices emanating from far off will be attenuated and softened by transmission through the atmosphere. This softening is due to the loss of fine details in the sound wave, particularly transients and what I'd describe as textural components. This leaves the more tonal characteristics of a sound, those we'd associate with pitch and frequency, to carry further. If you get the opportunity, such as the same species vocalising near and far, notice how a particular sound mutes, is smoothed, and becomes more tonal with distance. Rather than this degradation of sound being a hindrance to communication, many species exploit the phenomenon in their calls, something I'll come to in the next chapter.

Meanwhile, this consideration of tonal and textural brings our attention to the quality of sounds. Can you hear a contrast between smooth sounds, such as the melodic whistle of a songbird, and harsh ones like the buzzing of cicadas? Some may fall ambiguously between, perhaps the trilling of a cricket. Other species, birds in particular, may use combinations of raspy and sonorous elements. Perhaps you can find words to describe these abstract textures and qualities? They don't have to be accurate, simply terms that have meaning for you. Whatever imaginative adjectives you come up with, you'll be developing a personal vocabulary for natural sounds.

As you take stock of these diverse sounds, become aware of your emotional responses. Are there any sounds you particularly like or dislike? It is easy to enjoy melodic birdsongs or curious, quirky voices. But what of more abrasive sounds? An amusing social media post caught my attention recently, contrasting the restful dawn birdsong of Europe with an Australian burying their head in a pillow at the raucous sounds of cockatoos and kookaburras. Yes, listening is emotional. However if you attend more closely to those sounds you at first dislike, you may find yourself developing some empathy with the vocalising species. Personally, I've come to enjoy harsh and grating birdcalls, as they frequently embody character – the raspy, querulous calls of Apostlebirds are a delightful example. 🐦 By seeking positive interpretations, one can transcend unpleasant impressions to arrive at an empathic appreciation of all creatures. Being aware of your initial feelings is a starting point to finding a deeper connection.

Now, step back from individual sounds and take in the soundscape again. Can you hear 'layers' within the whole? This impression may be given by higher or lower frequencies, softer or sharper textures, distance, or the constancy or intermittence of voices. You don't need to analyse this particularly, just notice it as an aesthetic impression.

All the while, resist the temptation to turn your head overly. Rather than pivoting to face sounds, just allow them to come to you from any direction. This will enhance your sense of being in a place, and aid in calming the mind to become aware of small details.

And this is what we can focus on next. Are there any quiet sounds you've not noticed so far? Nature is full of subtleties that evade obvious awareness and are easy to miss. It is not only quietness of sound, but delicateness as well. As creatures vocalise, listen closely for nuances, hesitations, or tiny, expressive variations. Even when there are loud sounds present, there will often be moments of stillness between them – micro silences. It is worth listening for them.

In pursuing the small, soft and transient, you may eventually become aware that it's not all about the presence of sound. Quietness can be tangible in nature. I find that enjoying stillness is the key to maintaining attention over periods of time. At first, you may find yourself drifting off into distraction, to only be brought back by something that grabs your attention. But in time, if you keep listening for extensive periods, you'll develop the ability to transcend boredom thresholds. Maintaining awareness in the presence of minimal stimulation is a significant skill – a natural way of cultivating mindfulness. The reward is both a personal calm and a noticing of elusive and ephemeral happenings which may otherwise pass you by.

Learning Nature's Languages

All this focuses our perceptual awareness, and hopefully by now your ears will be out on stalks. You'll also probably be asking: what species is making that particular call? This is the question that represents curiosity kicking in.

In the days after hearing my mystery songbird on that ridgetop at Mutawintji, I still had no idea what species had so enchanted me. As Sarah and I hiked the landscape each day, I kept an ear open for a hint of that fluting call again. Then, one warm afternoon, I noticed piping notes coming from the sky overhead, and caught a glimpse of a Spiny-cheeked Honeyeater diving down into the crown of a tree. It turns out this was the bird's songflight, given in a voice unlike the Spiny-cheek's usual daytime repertoire, but recognisably akin to their dawnsongs – that part of their singing behaviour only heard in the predawn hour.

In retrospect, I realise that when I began identifying birdsong, I already had a reasonable knowledge of birds. All I had to do was associate unfamiliar sounds with species I already knew. Which made the exercise of connecting sounds to species enjoyably easy. With frogs, insects and other creatures however, I've had to start from scratch. So please don't be daunted – listening is such a rewarding way of becoming a naturalist.

No matter your level of natural history experience, you'll likely have some knowledge to begin recognising species by their voice alone. You'll probably realise you know more than you thought you did. But there'll still be many sounds you won't be able to pin down at first. In this case, the most conclusive and enjoyable way is to track the vocalising creature and find out who they are. First, if you are able to, make a 'quick and dirty' recording with whatever you have with you, a phone perhaps. This will capture the sound for later identification, if necessary.

Failing that (or in addition to), try using onomatopoeic words to describe the sound. This works well for many birdsongs and frog calls, and is most applicable with species that give simple, repeated calls. An onomatopoeic word is one that mimics the sound itself, and I'm not suggesting it be a 'dictionary' word at all. Instead, listen closely to the phrasing and structure of a sound, and try to capture it in your own made up words. This creative exercise will both aid memory, and develop your own very personal sonic vocabulary.

I'll offer an example from a species that inhabits the bushland around our home: the White-eared Honeyeater. They give quick, two-note calls with a 'choppy' quality. 🐦 Listening closely, I hear each note is bi-syllabic, so the whole call could be described as *whi-choo, whi-choo*, with the 'whi' component flicking up in pitch while the 'choo' is lower. The 'whi' is also accented, so I could refine it as *WHI-choo, WHI-choo*. And there is another subtlety; the 'whi' note begins with a sharp attack, *kWHI-choo, kWHI-choo*. To get a better sense of how it sounds, instead of the spoken voice, try it in a half whistle, half whisper. In devising similar mnemonics, you will remember the calls of your own 'kWHI-choo' birds. You'll also be continuing a tradition of naming creatures by their sounds that has existed since the dawn of humankind.

Of course you can explore a variety of other ways of remembering sounds, depending on the inclination of your listening mind. The call of the Pallid Cuckoo for instance could be thought of subjectively as sharp, loud and far-carrying, analytically as a quickly delivered series of around a dozen piping notes rising incrementally in pitch, behaviourally as oftentimes incessant, or characterfully as belonging to the 'brainfever bird'. 🐦

Of course, your conceptions of sound can draw on any associations you make – the more colourful the better. I recall the young son of a friend who told me he particularly liked the bird that sounded like a water droplet going 'Plop'! I was puzzled until I realised he was referring to the 'whipcrack' of the Eastern Whipbird. 🐦 Not knowing the familiar association, he heard it differently, in his own way. I could equally imagine our 'kWHI-choos' as the sneezing birds. So let your imagination freely associate, it's a playful way of recognising and remembering sounds.

Once you've documented your mystery sound – either by recording or memory – set off carefully, following the sound to eyeball who is making it, if possible. Nothing will reinforce your experience of this creature and its call better than both hearing and seeing it. In the process, you'll find yourself developing skills in how to quietly approach the subject of your interest without disturbing it.

If you don't manage to see it, and need to identify later, you'll at least have your recording or imaginative mnemonic as a reference. Seek out sound files that you can reference. When I began (back in the dark ages), these were best found on specialised identification tapes or CDs. Now there are many repositories of natural sounds in the digital realm. A little searching online and in app stores should bring up resources for species in your bioregion – frogs,[8] cicadas,[9] birds … For instance, the website xeno-canto has been created by a global community of enthusiasts, and presents recordings of pretty much every bird species on the planet.[10] The Cornell Lab of Ornithology is another sound repository with their own interpretive resources, such as eBird[11] and Merlin. Meanwhile, those early CD recordings have now been superseded by multimedia field guide apps that can be downloaded to your device.[12]

These resources, as comprehensive as they can be, may still prove overwhelming and frustrating for the beginner. Which of the many species possibly found in a particular location could that baffling sound be? Once again, here's the fun way of identifying it – ask someone. When I was growing up, local field naturalist societies were an invaluable source of information. With the advent of the internet, these small groups often struggle with membership, yet they remain a place to meet naturalists, some with a lifetime of experience, who will very likely welcome your curiosity with an open sharing of knowledge. So if you find yourself developing anything beyond a passing interest in the natural history of sound, seek out those who will share, enthuse and inform you. Nature is always local, and someone with local knowledge will likely be able to quickly identify that mystery sound and tell you more about the species. And if they can't, you can suspect you may have fortuitously documented something rare or unusual.

And don't give up on 'ye olde books' just yet. A good field guide will not only provide species information but usually include notes on vocal behaviour. The degree to which songs and calls are described and conjured up from the page will depend on the writer's sensitivity to sound. What you're looking for is an author who can convey sounds both accurately and evocatively – a poet naturalist.[13]

As you proceed in learning to identify species by their sound, a few generalities will become apparent. The first is obvious: every species has its own call, song or variety of sounds – its repertoire. This can vary a little with locality or race in some species (making identification from single recordings tricky), but should nevertheless remain reasonably distinctive. Even species with a wide diversity of calls, such as Australia's rosellas, often have a unique quality of voice which is recognisable with familiarity.

The second is that species of the same genus or family, while they'll each have their own repertoire, will often share something in common. Pigeons 'coo', parrots 'screech', thrushes such as Blackbirds sing melodically, often at dusk. As you go on, notice these commonalities, because they can give a clue as to what you're hearing. I find this especially useful when I travel to a new location, particularly overseas, and am confronted with an unfamiliar suite of species. I can sometimes make informed guesses based on those I already know. For instance, hearing a *"kek, kek, kek …"* may have me surmising that it's a local variety of kingfisher, as this is the type of vocalisation they often have.

Once again though, nature is full of surprises. Despite an expectation that one could at least categorise a sound as coming from a bird, frog, insect or animal, even at these fundamental levels, ambiguities may occasionally trip you up. Grasshopper Warblers are named for their insect-like songs, 🐦 several species of Asian green pigeons give delightful, soft whistles that are completely un-pigeon-like, 🐦 and the drumming of small woodpeckers can resemble a door creaking. 🐦 In the rainforests of the Solomon Islands, I heard a mammal-like sound – a half growling, half hissing coming from up in the canopy – and found it to be a Buff-breasted Coucal, a large cuckoo. 🐦

Creatures of Sound

As you listen and observe, it will become apparent that specific vocalisations are related to behaviours. Here listening opens us to the activity of the natural world and takes us into the communicative life of creatures. This is a massive topic, a scientific field known as bioacoustics. Each species may constitute a study unto itself, with its vocal behaviours possibly providing a researcher with years of enquiry. So while I'll touch on the breadth of communication in nature, anything comprehensive is well beyond my intention here. Instead, I'll focus my discussion on birds. They are familiar to everyone, and probably the creatures I have been most fascinated with myself. Their sophisticated acoustic behaviours will also allow us to explore the processes by which living systems function, offering us much to reflect on as we proceed.

We may think of birds as being creatures of the air, yet not all birds fly. All birds vocalise however, and they do so with a unique vocal organ, known as the syrinx. Comprising two vocal tracts controlled by groups of muscles, and situated where the bronchial tubes join the trachea, a bird's syrinx is remarkably flexible and actually capable of creating two sounds simultaneously. Indeed, many songbirds use this ability to weave complex songs by combining independent sounds from each vocal tract. The variety of vocalisations that we hear in the avian world, from the beauty of birdsong and the cries of fowl, to the hoots of owls and the uncanny vocal mimicry of parrots and cockatoos, is due to this marvellous organ. Incidentally, New World vultures provide the exception that proves the rule – being the only birds that lack a syrinx, they nevertheless communicate quite adequately with hisses and throaty grunts.

Some birds also produce sounds by biomechanical means, such as the slapping of wings, bill clicks, stomping or drumming on tree trunks. These sounds can be as articulate as any vocalisation. For woodpeckers, the rate and intensity of drumming is characteristic of each species. 🐦 Among the Palm Cockatoos of northern Australia, who have developed the remarkable behaviour of using a custom-made stick to drum against tree branches, individuals have been observed to create their own patterns and tempos.

Another unusual non-vocal communication is produced by the vibration of tail feathers in the Common Snipe. Over a landscape of lakes and peatbogs in upland Sweden, I witnessed birds silhouetted against a moody dusk sky, engaging in display flights of brief, steep dives, during which tail feathers were spread wide to set them vibrating in the passing airstream, creating a surprisingly loud, throbbing sound. This behaviour is all the more remarkable when one considers the bird has to judge the intensity of its dive just right; too little velocity and the feathers won't vibrate, while too fast may damage them. 🐦

All this tells of the importance of sonic communications in the daily life of birds. So as much as flight is a wondrous ability, I think of birds as creatures of sound.

Chapter 3

Nature Tells us Stories

The bush track that begins at the back door of our home winds for over a kilometre through eucalypt woodland. Sarah and I live in a rural area of temperate southeast Australia, surrounded by open, dry sclerophyll forest. It is a healthy ecosystem, with a mixture of native tree species, mostly box and gum, and under them, a diverse understory of native grasses, wattles and, according to the season, orchids and wildflowers. It probably looks similar now to how it always has, with a good community of ancient trees that pre-date the widespread land clearing that came with European colonisation. Some of these 'elders' may be several hundred years old.

Sometimes I think we are custodians of this land, and at others, that it looks after us. We certainly feel privileged to live here.

I walk this bush track – a mere path threading between the trees – most days. As I do so, I am always listening, even if sublim-inally. For me, the sounds of the bush are like having stories told aloud. They are stories of the various ways that the birds of this forest use sound to live. Any story begins with the characters in it, so I am pretty familiar with the species common to our area and recognising them all by their vocalisations. But the action of a story is in what the characters do, and why. This is where sound offers us a narrative.

To begin understanding these bush tales and yarns, I'd like to take you for a morning walk along our track. As we amble, I can tell you the stories I hear. So let's go for a listening walk.

As we set out, we'll put our attention to what we can hear. It may be worth pausing for a moment to do this. At first it seems quiet, but soon we begin picking up bird calls coming from various directions. The voice of each is characteristic, allowing us to become aware of the various species around us.

As we resume walking, I'll pose an admittedly rhetorical question: what is a species? I find it useful to think of each as being a unique solution to the problem of how to live. I know this is not a robust scientific definition, but it does focus us on the different *ways* that creatures inhabit the world. Unique vocalisations are necessary, as they communicate identity, but as we've noted, the real story is in why species use particular sounds, and the functions these have evolved to fulfil.

Homeranging – The Scarlet Robin

Up ahead, I can hear the delicate calling of a species that will allow us to begin considering this. It is a single, thin, quavering whistle, given every ten seconds or so. It comes from a Scarlet Robin, one of the *Petroica* genus of red-breasted Australasian robins, and one of the most beautifully plumaged birds of our woodlands. Following the song, we catch sight of the bird perched on a low branch – black back, white underparts and a vivid scarlet breast, its bill surmounted by a small but noticeable white forehead tuft. This is the male; the female is more demurely feathered, with a soft flush of salmon on her breast.

Now that we're close, we can hear the robin's song consists of a pattern of rippling sounds strung together like beads on a string. ◢ Making an onomatopoeia for it may give something like: *p-pi-pi-pri-tidi, prrri-tidi*. If this were human speech, we'd be recognising syllables and phonemes, but in birdsong, these fundamental sonic units are referred to as 'elements'. The robin trips them off with great fluency, the whole song phrase lasts only two seconds.

Our robins, like many songbirds, establish and maintain their living space by singing. The song we're hearing defines the area in our woodland where this robin is resident. You may be familiar with this being called a bird's 'territory', however I prefer the term 'homerange' as the colonial connotations of 'territory' and the implication of its 'defence' can colour our understanding of bird behaviours, a theme I will expand on later.[14]

For now though, we need to listen even more closely to this robin's song, as there is a significant nuance hidden within it, a quick trill in the middle; *p-pi-pi-pri-tidi, pRRRI-tidi*. Once your ear picks it up, it's quite noticeable. You may even notice there is another quieter trill in the first part of the phrase; *p-pi-pi-PRI-tidi, pRRRI-tidi*. These tiny trills are only audible up close.

With these trilling elements in its song, the robin is using the physics of sound to spectacular advantage. We can understand that sound attenuates with distance according to an inverse square law, losing power exponentially rather than gradually and evenly. In addition to this loss of loudness is a smoothing of the waveform, in which rough-textured components and finer details of sound deteriorate relatively quickly. We may perceive this as a more rapid loss of high frequency information than lower frequencies.

We can be more specific though. As we've already touched on, one can think of sounds as having tonal or textural qualities. Tonal sounds are those you can recognise as having an identifiable pitch; melodic whistles, or chiming, fluting, piping, ringing notes for instance. In contrast, those of a more textural quality lack noticeable pitch; buzzing, hissing, ticking, and harsh, grainy, raspy or scratchy sounds. These textural components don't travel far, and soon become inaudible, the overall sound decaying into any residual tonality.

Our robin is using this physical property of sound in a very pragmatic way. While his quavering whistle may be heard over several hundred metres, the fragile sonic details of those trills are not. This makes the trill elements a reliable measure of distance. Thus a listening robin can readily judge how far away a singing neighbour is. This ability is only possible if each robin has a near identical song as a reference in memory, which indeed they do. That birdsong could function in this way was first suggested by American ornithologist Gene Morton in the 1980s. Known as the Ranging Hypothesis, it is a concept supported by subsequent research.[15]

In this way, Scarlet Robins achieve several aims with the one song; broadcasting their location to neighbours, suggesting the extent of their homerange, and giving them the means to determine relative distance. This singing behaviour serves a significant purpose. Rather than having to 'patrol boundaries', a bird can conserve energy by singing from the core of its homerange, while also minimising physical interactions that may end up escalating. When you think about it, this economy of energy and avoidance of potential conflict, all enabled by a deceptively simple song, is cause for wonder. That the phenomenon exists, and is essential to the way much birdsong functions, has implications I'll return to later. But for the moment, let's continue our walk.

Sonic Lasers – The Bronze-Cuckoo

Leaving our robin behind, still singing, we note how those subtle trills quickly degrade to imperceptibility, long before its voice can no longer be heard.

Meanwhile we're approaching another prominent singer, giving a series of clear, downward whistles, one after another. Tracking the sound, we eventually catch sight of a smallish bird perched high on an exposed branch. This is one of our cuckoos, a Horsfield's Bronze-Cuckoo. For a bird of its size, its voice is surprisingly strong.

Which it needs to be. Cuckoos are brood parasites of course, and so don't establish or maintain a homerange. So that's not why it is singing. They are also usually present in lower numbers than the species they exploit, which implies the cuckoo population is spread out thinly. To attract a mate, what they need to do is announce their presence, and do so in a voice that will carry across the landscape. A high perch ideally suits this purpose.

Here again, the specific sound that cuckoos utilise is specially adapted. Most cuckoos, including our Horsfield's bronze, give their 'presence' call with strong, whistled notes. I think of their voices as being the sonic equivalent of lasers. Lasers emit coherent light – amplified and comprising a very narrow frequency spectrum. This allows a minimal amount of energy to be channelled into a tight beam that will transmit over distance. This bronze-cuckoo is using an analogous principle. It is focusing the possible spectrum of sound frequencies down to a single, whistled pitch, and amplifying it by vocal resonance – a sonic laser.

It is aided by the physiology of the syrinx, which, unlike mammalian vocal cords near the top of the trachea, is situated deep in the body. In most birds, this is at the confluence of the two bronchial passages from the lungs, but in cuckoos, it is located even deeper, in the bronchial tubes themselves. This anatomical configuration is very effective at projecting sound, perhaps similarly to the way a water pump is most efficient when pushing from the lowest point. Then, with bill wide open and throat pulsing, it is shaping sound, resonating and amplifying it by supple manipulation of the tissues of the upper airway and mouth.

Many songbirds utilise similarly articulated, whistling notes in their songs. Being purely tonal sounds, they carry for surprising distances across the landscape. By doing so, birds like our bronze-cuckoo are making the most of what their tiny bodies are capable of.

Keeping in Contact – Thornbills and Sittellas

The bronze-cuckoo now flies purposefully off across the treetops, possibly to try broadcasting from a new location. Ahead, we can hear a hubbub of twittering activity, and approaching see a party of tiny birds, some hopping along the ground while others flit among shrubs or glean in the lower foliage. All the while, they utter between them an insistent, tinkling chatter: *pinka-pinka-pink, Pit! Pit!* 🐦

These are Buff-rumped Thornbills, who are almost always found in small flocks. Their vocalisations are contact calls to maintain connection as the group forages through the forest. Once again, the specific sound is adapted to the purpose. Unlike tonal songs intended to carry far, contact calls need to inform about exact location. To do this, birds use brief, transient sounds, often with an irregular, textural waveform. This type of sound gives the ear a lot of information with which to easily recognise direction.

Localising sounds is a complex cognitive process, and birds achieve it similarly to ourselves. There are three criteria the brain uses to identify where a sound is coming from; relative loudness in each ear, the muting of high frequencies on the 'off' side, and the relative difference in arrival time of soundwaves between the ears, known as the interaural time difference. This phase discrepancy, measuring fractions of a millisecond, is by far the most significant.

Textural sounds are optimal for detecting this tiny time difference. Because the brain has to compare and match the sounds arriving at each ear, a more jagged and 'rough' waveform makes this determination easier, as there are more unique irregularities to match up. It is impressive that our human brains can achieve this. That birds, with far smaller heads offering correspondingly minuscule phase differences, can do it with accuracy speaks of their neural adaptation to the world. Respect for thornbills.

Listening to their twinkling *pinka-pinkas,* we now notice another sound; a quiet, repeated *zzzzt, zzzzt.* It is pleasant, but surprisingly easy to overlook. If it were coming from the tree canopy, you'd suspect it to be a small cicada buzzing. But this sound is coming from near the forest floor, from the midst of the foraging thornbills. Watching closely, we see one bird following another, its bill opening slightly to utter *zzzzt, zzzzt* every few seconds. Its somewhat plainer plumage markings identify it as a young one, and this insect-like sound is a juvenile buff-rump's begging call. Like their contact calls, it is a complex, rough-textured sound, and very easy for a parent to localise – this little buffie is not going to go hungry.

There are more birds around now than only the thornbills. I'm also picking out a soft, high frequency *see, see ... swee ...,* and looking up into the canopy spy a mob of Varied Sittellas. 🐦 I do love sittellas. They remind me of the nuthatches of the northern hemisphere, being of similar size, shape and habits. Traveling in small flocks, they patrol along branches and trunks of trees, often downwards head first, or even hanging underneath on strongly-clawed feet, meticulously examining and probing the bark continually. Our antipodean sittellas are entirely unrelated to the northern nuthatches however – they are fine examples of separate branches of evolution converging on similar solutions to a universal problem, in this case a bird adapted to extracting insects from tree bark. They even both have slightly upturned bills ideal for chiselling out concealed morsels.

The sittellas' light voices are all around us, translucently high in pitch. Their contact calls have a different quality to the thornbills, but are also short snippets of sound, frequently given, a continual stream of thin twitters that allow each individual to forage independently, while maintaining awareness of where its companions are.

Actually it is more complex than just keeping in touch with companions, as these thornbills and sittellas are moving through the forest as one flock, and the sounds they create are keeping the whole group together. So they are not simply calling to their own kind, but listening to their companion species as well.

Now another bird appears, a Grey Fantail which darts out from the forest mid-storey on brief forays in pursuit of aerial insects. It is chaotically acrobatic, flicking wings and tail into a tumbling flight almost devoid of any directness. As it flutters hither and thither, it calls on the wing with tiny 'sneezing' sounds; *nyip … nyip …* Once again, this rough-textured sound makes an excellent contact vocalisation. As another fantail appears, these sneezes become more frequent, accelerating to tip into a delightful cascade of silvery whistles – a mini-carillon of tiny bells. This tonal utterance is the fantail's actual song, given to bond with a mate, so these two are likely a breeding pair.

Distracted by the antics of the fantails, we now notice the collective fuss of contact calls lessening. The thornbills and sittellas are drifting off, flying in twos and threes from one shrub and tree to another. As the flock recedes, they take their shared soundworld with them. For such small creatures, often well camouflaged and always on the move in a visually complex habitat, their contact calls are an efficient way for them to stay together and not lose anyone.

The Art of Ambiguity – Pigeons

By a happy co-incidence, I soon have the opportunity to illustrate the acoustics of directionality by contrasting with the opposite. There is a regular, almost subliminal booming coming from somewhere; a low *oooom* … uttered every few seconds or so. It is a Common Bronzewing, a pigeon. It seems some distance off, but exactly where is difficult to determine. I twist my head from side to side trying to get a clue as to its location.

The bronzewing's call is so ventriloquial because it is an almost pure sinewave. This means that the form of each wave cycle is smooth and similar to the next. Consequently, our brain struggles to detect that phase difference. It cannot easily match individual wave cycles from one ear with the other, and thus get a measure of the relative time delay. The result is that the bronzewing's call just seems to come from everywhere. Plus, it is of such low pitch that frequency muffling and loudness differences aren't significant either, so twisting my head isn't helping.

The bronzewing's call certainly travels well, it seems to permeate the air. But projection is not the primary reason it has these tonal characteristics. Think of it from the pigeon's perspective. It is a lovely, plump meal for a passing goshawk, sitting on a branch and conspicuously advertising its presence. If I were a pigeon, this would be cause for some anxiety. By having a call that is so difficult to localise however, our bronzewing is relatively safe, hiding in plain sight – or in this case, sound.

Ambiguity may have its benefits, but also poses a problem. How does the bronzewing's mate or neighbours know where it is? I think the answer has something to do with monotony. A bronzewing will call on and on, uttering an *oooom* every few seconds for an extended period. This allows time for fellow pigeons to hone in on its whereabouts. Once closer, subtle higher frequency components in the bronzewing's voice can assist in fine tuning its locale. A goshawk on the wing however, may only hear a call or two in passing, not enough to get a firm bead on its prey. If it settles close enough to listen, the pigeon will likely spot it and go quiet, or take to the air with a clattering of wings – thought to be an alarm signal among pigeons.

I'll tell you more about the low frequency calling of pigeons later, but for now, let's move on.

Social Connection – White-winged Choughs

Up ahead there is more activity; a commotion of squawking coming from a mob of largish, black birds strutting around on the ground. By a quick count, there may be around a dozen in this group. We can see them, patrolling back and forth, probing at the dirt with their bills, tossing aside leaf litter and sticks. They are White-winged Choughs, another uniquely Australian bird that – being black-plumaged, gregarious, terrestrial foragers with down-curved bills – resemble, but are unrelated to, the choughs of the northern hemisphere (more convergent evolution).

They're gradually moving in our direction, so we'll pause quietly here and see what happens. The juvenile in the group is particularly un-ignorable, continually giving calls which can be described as infinite variations on a thin scream. It seems pretty evident that this is another 'feed me' call.

Choughs are very obviously a social species, and their primary means of bonding is vocal. It is impossible to miss a group of them in the bush, as they are constantly in sonic contact. While their behaviour can be endearing, their vocalising may be less so. In addition to the youngster's begging squeals, the group give soft whistles of reassurance, murmurs of absent-minded contentment, plus a few harsh and grating expletives thrown in for good measure.

Whilst these vocalisations may not always be pleasing to the human listener, their rich sonic texture is beautifully adapted to the chough's requirements; they are very expressive sounds. Choughs have complex social lives, which are facilitated by the capacity of their vocalisations to convey personal identity and state of being. This expressiveness of voice is found in other social birds, animals and indeed, ourselves.

The choughs have now approached and suddenly become nervous of our presence. With loud, rasping squawks, they take alarm and fly up into the treetops. As they do so, we see the broad, white wing panels that give them their name. They gaze down with seeming indignation, and now commence a different and quite unexpected sound, a collective chorus of deliberate, down-gliding whistles – a cascading, sonic waterfall. For me, to hear a group of choughs uttering these glissandi of clear, sliding tones is one of the most attractive sounds of the Australian bush.

The choughs however are not intending them to be pretty. At first I interpreted this call as their alarm, but watching them sitting safely but warily, I've come to conclude they are asserting something along the lines of; 'if we stick together we'll be alright'. So it's a mix of agitation, caution, and social affirmation – more 'yellow alert' than red.

What I find interesting about this call is a nuance one may easily miss: each descending whistle is prefaced by a sharp bill click. It's obvious once you notice it. The click is significant, because like contact calls, it is a sharp, transient sound, and hence easy for the ear to detect directionality. By contrast, their whistles are pure notes, which carry further but are less directional. So in this one integrated sound, combining mechanical and vocal, the choughs are signalling their individual locations to each other, bonding as a community, and creating a collective ruckus that says they're not easily intimidated by intruders to their domain. I know they're not just objecting to us, as I've heard them respond similarly to prowling foxes and even in interactions with neighbouring groups of their own kind.

Voices of Intelligence – Cockatoos and Ravens

We've now arrived at the end of our bush path where it opens into semi-agricultural country. As we look out, a group of Sulphur-crested Cockatoos fly past, white against the blue sky, screeching to each other on the wing. 🐦 It's a big sound which tears at the air, and certainly carries well, but the significance of those screeches is in their sonic complexity. Each rasping vocalisation is spectrally 'messy'. Toneless and chaotic, they are nevertheless expressive, with a wealth of sonic detail across a broad frequency range. Cockatoos are intelligent creatures, and like choughs, highly social. No surprise then that their calls have similarities in acoustic complexity, as they too are expressive, communicating identity and facilitating their interactions with each other.

Turning now to retrace our steps homeward, we notice a pair of Australian Ravens flapping slowly overhead, calling as they go. Like many of their family, the corvids, their voices can sound very human-like. After a series of *ark, ark, ark*, passed back and forth between them, one bird ends with a mournful, drawn-out, descending, *aaaaaaaarrrk*. It's a sound that embodies something of a laconic Australian character – a droll, mournful sigh which is so emotive that it's difficult not to smile. 🐦

There is good reason for us to respond to ravens and crows in this way. Unlike much birdsong, which is composed of whistled notes or spectrally complex sounds, corvids produce sounds rich in harmonic overtones.

Harmonic overtones, sometimes called higher partials, are a consequence of the physics of sound. When a note of a certain pitch is produced, there exists the possibility of resonant notes at higher frequencies being simultaneously created. Due to the physics of sound production, these resonances occur in a fixed pattern, the frequency of each overtone being a successive whole number multiple of the fundamental frequency. The resulting sequence of higher partials is known as the harmonic series, and from its pitch intervals are derived the scales of many of the human world's musical cultures.

However, actually hearing overtones 'in the wild' is a little tricky, as they are hiding, quite literally, under our noses. They are a feature of mammalian voices, with our own being particularly rich in them. You'll be familiar with overtones, as they are the aspect of sound we manipulate in shaping vowel sounds. To get a sense of this, try humming a note with lips closed, thus suppressing the resonance of the voice. Now, staying on the same note, open to a flat 'O' vowel – 'Or' – and feel that additional 'ring' to your voice. Next try the flat 'A' – 'Aaaaa' – and an 'Eeeee'. While your fundamental sung note hasn't changed, the resonance of your voice has, from lower resonances in the throat, to higher ones formed at the back of the mouth, to those that tickle your sinuses. Each of these resonances is a note in itself, it's just that, like the rainbow of pure colours comprising sunlight, we don't hear overtones as distinct sounds, but blended into a kind of vowel soup. Nevertheless, it is the palette of harmonic overtones that adds colour and expression to the human voice, allowing for nuanced communication.

Returning to birdsong, it is interesting that the communicative requirements of many species are achieved through pure-toned whistles that have very little overtone content. Crows and ravens are among the few songbirds whose voices are rich in overtones, and I suspect that they play a similar role for them as ourselves. By adding colour to the voice, they support a different kind of communication; less precise than much birdsong, but more expressive. Corvids are renowned as highly intelligent creatures, with complex social interactions. For them, the more soulful opportunities of harmonically rich vocalisations are appropriate to their intelligence. We relate to the idea of them 'talking', because in their own unknowable way, I believe they are.

Intimacy – Fairy-wrens and Robins

Presently, we come across a party of Superb Fairy-wrens in the undergrowth by the side of the track, announcing their presence with animated contact calling. They are keeping close together as a family unit, and as they dart around among low shrubs and forage on the ground – little balls of feathers, tails cocked and twitching – they give a gentle stream of brisk *prip, prip* calls. No problem hearing exactly where each bird is, as they're classic contact sounds; short and brittle. As long as it's safe, they'll call to keep their group together.

As we blunder past, their *prips* become a little more agitated and they flutter for safety into denser vegetation. But they seem to know we're no significant danger. The moment there's a real predator, they'll change to a high-pitched *seee, seee, seee …* in alarm. Although it's not as apparent as with the bronzewing, those *seees* are also pure sinewaves, and on the relatively rare occasion I've heard them, they seem to hang disembodied in the air. Fairy-wrens are not alone in this, with many of our smaller birds such as thornbills using similarly ventriloquial warning vocalisations, allowing them to sound the alarm while remaining difficult to locate.

The fairy-wrens have quickly got used to our presence and return to their usual activities. As we listen quietly, they settle into making soft vocalisations among themselves. This combination of squeaks, high-pitched whinnies and little trills is very endearing. It is one of my favourite bush conversations to overhear. These birds are simply happy, contentedly twittering to each other in proximity. All is right in their world.

Continuing to retrace our steps, we are approaching the place on the track where we heard the Scarlet Robin previously. He has stopped singing now, but as we draw near we notice a quiet, ticking sound. It is curious, not at all bird-like, yet there is our male robin, perched on a low branch, emitting an irregular stream of ticks and rattles that one would be struggling to hear much more than twenty metres away.

But it is not intended for public dissemination, as becomes obvious when we spot the female perched on an adjacent branch close by. He darts to another perch, and she immediately follows. When they settle close to each other, the stream of ticking calls intensifies. It seems the female is also contributing her own quiet sounds. I've heard Scarlet Robins making a similar call, perhaps slightly louder, when it seems to be a contact vocalisation. But now, it appears to be more associated with bonding, an intimate vocalisation between the two birds. Perhaps we're hearing sweet nothings in robin talk. Once again, this discreet vocalisation is superbly appropriate for the purpose.

The robins soon move further off, still consorting with each other. Leaving our lovebirds, we resume walking. It is now later in the morning, and the bush is quietening down.

Family Dynamics – Magpies

We've nearly returned when the sound of wings comes from above. I smile without looking up, immediately recognising who it is. The primary wing feathers of Australian Magpies are stiff, and with their pattern of determined wingbeats make for a distinctive sound in flight. Sure enough, two magpies fly overhead to alight in the crown of a tree a little in front of us. A third has been following and joins them. I must admit, I've been hoping we'd hear the lovely singing of magpies and get to observe their family interactions. We'll carefully move a little closer – our wild magpies are not as trusting as those that live in more urban settings.

Perched near each other in the morning sunlight, we see one tilt its head skywards and begin uttering a melodious stream of bubbling warbles and rippling notes. The other two are poised and join in, adding their voices to collectively weave a song in which their individual contributions are indistinguishable. Melodic and pleasing to the ear, it is easy to understand why the carolling of Australia's magpies is so affectionately admired.

Australian Magpies are another species that live in family groups bound by complex social relationships. They play, squabble, learn, help each other, hang out and forage together. Singing as a family ensemble is a characteristic behaviour they share with butcherbirds, to whom they're closely related.[16]

Magpie song is not all soft and melodious though. If you listen closely, you'll often notice that embedded within their warbling songs are some loud and strident elements. I call these 'rawking' syllables, and they consist of hard, rapid trills. While having a somewhat harsh texture in proximity, my guess is that's not how they're intended to be heard. The magpie's forceful rawking notes travel over the landscape, often being the only component of their song audible from a distance. 🐦 These far-carrying sounds seem intended for the ears of neighbours because, being trills like those encoded into the Scarlet Robin's song, they will allow adjacent magpie families to discern distance.

Thus they are another example of a single utterance having multiple purposes – and audiences. Within the complex warbling that fulfils social functions among a family group, magpies simultaneously create signals to communicate with their next door neighbours. Interestingly though, relative to the robins, their more social lifestyles have resulted in a reversal of sonic form and function. While the robin discretely embeds its ranging cues within its song, the magpie's 'rawks' are pumped out to far-flung neighbours, while their limpid songs affirm the intimate bonds of family relationships.

There is a further aspect to this phenomenon. Across the continent, magpies fall into several races which display regional song variations. In arid, inland Australia, magpie voices can be downright raunchy – I've encountered birds whose pleasant warbling concluded with a rising, rawking inflection that hit my ears like a whiplash. 🐦 I suspect an explanation for this may be that less fertile habitats require larger homeranges, and hence more strident calling. The implication is that the structure of magpie song is adaptable, and tuned to the size of the homeranges the birds occupy.

Another feature of magpie song is that individual birds (likely the male of the family) can often be heard singing at night. It is lovely to hear them do this, because they do so in a soft and introspective voice as though singing to themselves. 🐦 Whilst magpies will warble quietly in the dark wherever they occur, I've found outback birds to be more frequently heard doing so, with a unique nocturnal repertoire not given during the day and lacking any rawking elements. I'd love to know the significance of this.

Sounds Fit for Purpose

In our short walk, we've met a few of our local species. Among them, they've displayed some of the many ways birds use sound to survive. They've vocalised to maintain living space, stay in contact, feed together, defend themselves, announce their presence, communicate emotional state, alert each other, facilitate their social relationships, establish communities, bond, be intimate, and for pure pleasure. We can relate to how pivotal these functions are to life, and so get a sense of just how crucial sonic communication is for them.

We've also made connections between the sounds employed by species and how wonderfully adapted they are to their purpose. The vocalisations of each species are bespoke, utilising the physics of sound to encode precisely the required information. That every species has its own repertoire reflects that each have a unique identity and communicative needs. All soniferous species – those that communicate acoustically – have life stories expressed in their vocalisations.

What I've discussed here is only the tip of a single head feather of the penguin sitting on the proverbial iceberg. I hope that, through the species that live around me, you're getting a sense of how to listen to those around you, and make connections between sound, acoustics, individual creatures, their habitat, behaviours, daily existence and communicative purposes. With this insight into the role of sound in nature, we can begin to appreciate how listening can take us into the lives of other beings.

The Tall Forests

We were camped out in the temperate forests of far southeastern Australia – and the weather was turning bad.

In the few weeks that Sarah and I had been in the upland forests of far East Gippsland, we'd found them to be a realm of sheltered gullies and birdsong, where morning sunlight shafted between towering trees. Mountain Ash, *Eucalyptus regnans*, the Earth's tallest flowering plant, dominated, along with Candlebarks, Brown Barrels and Shining Gums. Their stately trunks, smooth mottled in yellows, greys and soft ochres, created an impression of cathedral columns ascending to support a canopy some 60–80 metres overhead. Below them, an understory of giant tree ferns, banksias, blanket leaf and silver wattles were often so tangled as to be virtually impenetrable. Pockets of cool temperate rainforest, often protected in valleys and threaded with rushing streams, were remnants of ancient Gondwanan ecosystems. Occasionally, we would come across flowers of the Gippsland Waratah, bright red and the size of a hand, held palm and fingers upward, glowing in the forest depths.

The whole place was not only majestic to the senses, but ecologically precious. However it was under ongoing threat of clearfell logging, and conservationists were attempting to raise awareness of the region's biodiversity values – which is why we'd chosen to come here for our second sound recording project.

A few days previously, we'd driven up a remote track to arrive at a secluded location known as Waratah Flat. Exploring around our camp, we found the forest full of birdsong. The cracks of Eastern Whipbirds ricocheted through the trees, the voices of Gang-gang

Cockatoos sounded like creaking branches as they took lazy wing overhead, 🐦 whistlers sang like audible sunshine, and honeyeaters chipped and twittered in gregarious abundance. 🐦

But now the forest had become anything but peaceful, as a massive storm system blew in from the southwest. Ferocious winds lashed the crowns of the trees and clouds scudded low overhead, shrouding everything in mist, rain, or swirls of sleet and snow. Every now and then we'd hear the sickening groan of a tree branch tearing loose, and the heart-stopping crash of it hitting the ground. To attempt to drive out along the kilometres of forest tracks would have been foolhardy. The forest was being shredded, and we were huddled inside a small tent, waiting for the next branch to be tortured loose and come shattering down, unable to do anything but sit out the storm. 🐦

It seemed an opportune time to consider our future.

This was a year after our sojourn at Mutawintji, and Sarah and I were on the cusp of deciding to redirect our efforts away from music toward the publishing of natural soundscapes. So far on the trip, we'd collected some wonderful recordings. The bird-song we were documenting was full of character – the cries of Black-Cockatoos, 🐦 delicate twitters of Rose Robins, and the mimicry of Lyrebirds. As these voices of the forest permeated us, we increasingly came to feel that they should be heard in their own right, without any human or musical overlay.

As the gale tore at the forest, we came to our decision. At the time, very few people were making nature recordings available commercially, and we felt we were embarking on a project with an uncertain outcome. We would be going off the beaten track, in more ways than one, choosing a life where we'd get a real tan rather than a studio tan. Yet we felt that, if any two people were the ones to create and publish nature soundscapes, it was us. We both had a sense that this was 'our thing to do', and our label's previously chosen name, Listening Earth, became unexpectedly prescient.

*

Once the storm had passed over Waratah Flat, I began wandering the forest at night, hoping to record one of the animals that had drawn us here; the Yellow-bellied Glider.

Yellow-bellies are native, nocturnal mammals, about the size of a cat, with a love of feeding on sweet tree sap. To move around their feeding trees, they climb to a high position and launch themselves into the air, gliding down on a parachute of outstretched skin membrane between arms and legs, their furry tails acting as rudder, to alight on a neighbouring tree. As they do so, they give a blood-curdling, grizzly growl. We'd first heard about them from our friend and ecologist Steve Craig, who described their vocalisations as somewhere between the biggest tummy rumble you've ever had, and the sound of a cat being swung around by its tail. Who could resist such a description?

Over several nights, I'd heard yellow-bellies in the distance, but hadn't been lucky enough to encounter one close by. It was not easy finding them. Being utterly dark in the forest, with starlight only occasionally visible through the canopy overhead, catching sight of one sailing through the trees was unlikely. So instead I was listening for them, hoping to detect their scratching of claws on bark.

Eventually, it was a soft moan above which alerted me to the presence of one. Pointing my microphones upward, I waited. Now came some scrabbling as the animal climbed higher, and another quiet moan. Then, suddenly, that exhilarating call signalled its leap into the air, and traced its descending glide between the trees, terminating as the animal landed a little further away. Just as Steve had described, hearing a yellow-belly in its growly flight was certainly worth the effort of coming here.

Over the following weeks, I collected hours and hours of recordings, representing many of the creatures native to these forests. Yet there was one we never heard, which was disappointing, as it was an iconic species; the Sooty Owl.

Sootys are rare. Inhabiting only the thickest rainforest gullies in eastern Australia, their numbers are not great. Named for their ash grey plumage speckled with paler markings, they are *Tyto* owls; that genus recognised by their heart-shaped faces. In the dead of night, sootys hunt on silent wings, using an unerringly accurate sense of hearing to locate their prey. The only sign of their presence is likely to be their distinctive call; a shrieking, downward whistle often described as akin to a 'falling bomb'. But in all our weeks in the forest, and despite listening most nights, we'd not heard a single one.

Then, in the concluding days of our field work, a local naturalist gave us a suggestion of where we may, if we were lucky, locate a Sooty Owl. Along a forest track that led up the valley of Ellery Creek, Sarah and I set off on one final foray. Massive tree buttresses loomed around us in the darkness, and the nocturnal forest was still and quiet. Stopping every now and then, we listened out into the night.

After several hours, we'd walked kilometres, but not heard much at all. Feeling a little deflated, we decided to call it quits and return. With feet crunching on the leaf litter, I suddenly stopped and froze. I'd heard something, far off. We stood listening intently. Nothing. Then – there it was again, this time closer; a half shriek, half whistle, falling. It had to be a sooty. Out came the mics, and I switched on, just in time to catch another call, plus a responding one; two birds, both now closer.

There are moments when hand-holding microphones (as I was then), that I wished I didn't get quite so excited. I was trying not to tremble the highly sensitive mics, because, as elating as it was to capture the sooty's contact calls, what happened next was utterly serendipitous.

Unseen, both birds must have flown in and landed high in a tree almost directly overhead, and then began vocalising with the most delicate, trilling calls. My first impression was of how unusual these sounds were for an animal to be making. They reminded me of a mechanism whirring, revving up and subsiding, yet the expression and delicacy of their song was so alive. The two owls kept trilling back and forth to each other for some time, maybe twenty minutes, stopping as abruptly as they began. We caught a last, furtive hint of a dark shape moving off against the sky, before quiet descended again. Switching off the recorder, we hugged with delight. We'd just been privileged to overhear a pair of sootys sharing their tender, nuptial song.

With field recording complete, 'Tall Forest' became the first purely nature CD we published. In the years and decades following, our catalogue of recordings grew as we travelled widely, documenting a range of wild environments, firstly in our home country, and then increasingly from around the world. Yet those forests of southeast Australia, with their cathedrals of trees and alive with birdsong, were the habitats that first inspired us.

Chapter 4

Hearing Sentience

Torresian Crows

It was fairly obvious that the local crows felt they owned the place.

In the central Australian desert, we'd set up camp for a few days, and the next morning watched in amusement as a mob of Torresian Crows arrived to investigate our vehicle. You could see their curiosity as they took in this new object in their domain, walking around and under our campervan, inspecting it closely, hopping up on the bumper and craning their necks to scrutinise the radiator. The windscreen wipers were pulled back and released with a satisfying snap. Once on the roof, they first took turns swinging off the roof rack, before discovering that the real fun was in sliding bodily down the windscreen. They'd do this repeatedly, flapping back up to take another turn amidst little caws of delight.

Crows and ravens are known for their intelligence. They have keen memories and display problem-solving abilities that have turned them into social media celebrities. Some species even make and utilise tools. As we've noted earlier, their vocalisations are perfectly suited to expressiveness and sociality. I've also been told of crows singing a rain song together, signalling impending stormy weather sometimes up to two days before it arrives, and more reliably than the bureau forecast.

Over the following mornings, I recorded in the surrounding mulga woodlands, placing my microphones on a tripod before sunrise, and leaving my equipment unattended to passively record until late morning. This has become my favoured way of recording, as wildlife will behave naturally and come much closer to the microphones without my presence.

Listening back later, I found that on many occasions, my recording gear, sitting out alone in the landscape, had been visited by inquisitive crows. On headphones, I could hear a bird flying in and landing, a shuffling of wings and their guttural voices at close range. Remembering where I'd put my equipment, I could sometimes recall the actual bush they must have landed in.

On one of these recordings, a bird did something unexpected. Landing on the ground a few metres away, it gave a few calls, before hopping toward the microphone. In my mind's eye, I could envisage it clearly. It came to stand right below the tripod, and began to make quiet and intimate vocalisations that I can describe as muttering and musing to itself. They were soft gurgles and aspirated moans, of the kind we may accompany a thought process of "hmmm, now, let's see, what's going on here?" Then it gave a half sigh, half gargle, followed by two louder calls, one a higher note than the other; "Oh, well". There seemed an air of disappointment in its voice. It then gave a few louder calls, presumably to its nearby kin, as if to say, "nothing to see, moving along ...", before flying off.

Removing my headphones, I was aware that I'd just heard another mind.

Until that moment, I'd been interpreting nature's sounds in terms of species, their vocal repertoires and behavioural associations. I hadn't been connecting with the living creature itself. Like a picture coming into sharp focus, I realised that all along I'd been hearing consciousness – the sentience of other beings.

That curious crow has had a transformative influence on me, and I've played my recording of it when giving public talks over the years. On one occasion, an audience member asked: "Aren't you just anthropomorphising?" This is an important question, so let's explore it.

When we communicate with each other, we do so on the basis of a kinship of mind, a shared human experience. When we turn our awareness beyond our species however, we unthinkingly draw a line, beyond which exists life which is not us. In the companionship of pets or tame animals, we may blur that distinction, affectionately shepherding them into our realm and talking to them as members of our own family. But for wild creatures, we often assume their consciousness is way too alien for us to really relate to.

For me, recognising this hidden assumption and putting it aside has opened a whole new dimension of listening. Its essence is empathy and its outcome is a deeper connection. As I've suggested, listening is a powerful means of venturing beyond the anthroposphere, our human bubble. Through listening, we can relate to the qualities of a wild creature's voice, and sense the emotion expressed. Empathic listening can allow us a real perception of another creature's life which, while still foreign, is not so alien that you can't feel your resonance with it. Call it instinct if you prefer, but it is no less legitimate for that. The more you familiarise yourself with an animal, and listen to nature generally, the more revealing this way of perceiving becomes.

Empathy thus gives us a valid knowing of other creatures. This has come to be acknowledged in animal behaviour studies. Famously, primate researchers Jane Goodall and Dian Fossey broke the rules of scientific objectivity in the study of wild animals, instead building relationships with them to understand their lives. Research has since shown the extent of emotion, individuality and agency in animals – even empathy is recognised. This observational data has been complemented with neurochemical studies, demonstrating that animals experience pleasure, fear and curiosity just as we do, and likely for similar evolutionary purposes. Even a comprehension of selfhood, what is termed 'theory of mind', is increasingly being suspected as not solely a human capacity. I have little doubt that our inquisitive crows had a pretty good sense of themselves.

At the heart of this is not so much how other creatures may experience the world, as how we do. By dismissing our innate sensitivities as mere anthropomorphising, we hinder our ability to comprehend the living world. We diminish our own perceptive capacities and compromise our potential to acknowledge the rich emotional lives of animals. By objectifying other creatures, we put ourselves at a distance from them. For me, this separation is simply an expression of Western culture's prevalent disconnection from nature. And when you think of it, objectivity itself is equally a form of anthropomorphising – projecting our 'human form' – but from an alienated rather than connected state.

The naturalist and writer Carl Safina, in books such as *Beyond Words*,[17] speaks of levels of empathy between ourselves and wild creatures. First is the recognition of similarity when we encounter another living being. Next, a sympathetic response that acknowledges a shared commonality of animal feeling and experience of life. And then, compassion: putting those realisations into action.

Empathy is a necessary antidote, not only to personal separation, but ultimately to our domination and exploitation of nature. Through empathic listening, we can touch our kinship with another being. When we do so, we extend our 'moral circle'; the reach of our consideration of others. Empathy is one of the necessary expanders of our moral circle, as the degree to which we recognise the minds of other animals informs our concern for their wellbeing. This is why listening is such a radical act; it promotes greater compassion and care.

Now, when listening, I still identify species by voice and associate their calls with behaviours. I still hear those stories of nature being told aloud. At the same time, I'm also aware that each voice I hear is an expression of a kindred mind, alive to the world in ways I'll never truly know, but nevertheless can sense.

Instead of hearing a soundscape, I feel myself a listening participant in an audible mindscape.

The Nightingale

The song of the Nightingale has been celebrated in European culture throughout the centuries. Shakespeare repeatedly eulogised its sweetness of voice, and Beethoven appointed a flute to mimic its song in his pastoral symphony. Milton wrote of the Nightingale as the 'most musical' bird, and hearing it inspired both Coleridge and Keats to reflect on nature and human existence. In his 'Ode to a Nightingale', Keats describes one singing with 'full-throated ease'.

In 1900, the voice of a Nightingale, documented on an Edison wax cylinder by Cherry Kearton in a garden in Kenley, England, became the first recording of wild birdsong. Ten years later, the earliest gramophone record of birdsong featured a Nightingale recorded by Karl Reich in Berlin. The composer Respighi, in what was a pioneering and controversial use of electronic sound in music in the 1920s, called for a recording of Nightingale song to be played through a loudspeaker placed among the orchestra during performances of his tone poem, 'The Pines of Rome'.

Considering this adulation, I have to admit that on first hearing this renowned bird for myself, I was a little disappointed. It sounded scratchy and twittery to me. Is this what all the fuss was about? In my defence, I suggest that if those European writers and musicians had heard the sublime phrases of Australia's Pied Butcherbirds, the wistful melodies of India's Malabar Whistling Thrush, or the cheery cadences of Asia's bulbuls, they may have switched allegiances. Or, as I came to appreciate after listening more closely to the Nightingale's song, perhaps not.

I had the opportunity of really spending time with Nightingales in a small copse of trees, hidden away among the fields of agricultural Turkey. For miles around, the landscape showed evidence of having been cultivated since antiquity. We spent some time exploring the district for a place to record nature relatively undisturbed, eventually discovering a fold in the hills, through which flowed a small stream overgrown with willows, alder and berry brambles. Adjacent, an abandoned orchard had gone wild. The copse was its own microhabitat, and here, Nightingales had found a home.

As their name suggests, Nightingales often sing in the dark. Listening to one in the stillness of night serenading from deep in a tangle of vegetation, I heard it pour forth a continual stream of phrases, one after another, each different to the next. I had to admit, this bird was a seriously accomplished vocalist. My growing appreciation was reinforced when a fervent but rather tuneless rendition of the predawn call to prayer was broadcast across the landscape from a nearby village. 🐦 This reminded me of the tale of Saint Francis of Assisi, who reputedly sung back and forth with a Nightingale all night, eventually conceding the bird was the better.

As I listened to the Nightingale continue its performance, I tried to follow its progression of song. Each of its phrases was short; a compact collection of chips, staccato twitters, trills or rippling clicks, each separated by a short pause. Its repertoire was diverse and ever-changing, but among all the variety, a single phrase caught my ear. It began as a rapid series of whistled notes on one pitch, ever so gradually slowing and dropping in frequency. It just kept going, slower and slower, lower and lower, teasing out the notes over half minute, until … *"pip, pip!"* – it concluded with a double full stop. I smiled – this bird really was playing with sound. 🐦

As with the crow, I've played this recording to many audiences, and every time, it raises a ripple of surprise and amusement. It is a spontaneous response, which I consider a delightful example of cross-species, human to bird empathy.

Considering the extended repertoire of the Nightingale, comprising these multitude of separate phrases, I am drawn to wondering whether creativity is involved. Is this bird simply tripping off a collection of stock phrases, or is it personalising them in some way? Is the sequencing of phrases significant? Is it spontaneously improvising? Not being deeply familiar with Nightingales, I can't answer these questions, although I'm confident the bird's mind is actively shaping its song. In its spirited variety of song phrases, we can follow the Nightingale's mind at play.

In suggesting this, I'm reflecting on two Australian species that have told me what the avian singing brain is capable of.

Pied Butcherbirds

O that those European bards had heard Australia's Pied Butcherbirds! In this alternate reality, they would perhaps have penned prose celebrating the melodic purity of its song, or extolled the union of male and female voices in creating a duet so skilfully entwined as to be sonically seamless. Despite this loss to literature, contemporary Australians do recognise the Pied Butcherbird as a beautiful voice of our drier inland and subtropical regions.

During the day, a pair of birds (or a family group) will sing together, one beginning a phrase and the other completing it, the passing of the melody between them being so flawless as to seem like one voice. One often has to listen and watch closely to recognise this is actually happening.

What is not so commonly realised, is that while male and female birds duet exquisitely during the day, at night the male will sing solo. And what a song! If the Pied Butcherbird's diurnal song is extrovert and precise, it's nocturnal song is slow and reflective, as though given while half asleep. Reverberating off the landscape in the still air of an outback night, it can be sublime.

However recording it does require getting up out of cosy bedding into the crisp, desert night, and blundering round the scrub in the dark. Having done this on numerous occasions, I can say it is always worth it. It has given me some of my most memorable recordings, with one in particular, of a bird singing in the cathedral ambience of Ormiston Gorge in the Western MacDonnell Ranges, having inspired several musicians to compose their own responses.

A Pied Butcherbird's night-time song is made up of one melodic phrase after another, each spaced with a relaxed pause. As time goes on, one notices certain phrases recurring. They may be sung precisely the same, or ornamented with a little extra trill or grace note. As the bird continues, core phrases can be recognised, to which he returns regularly.

Fascinated by this, I became curious as to the sequencing of these song phrases. Which one followed the previous? Was there a pattern? Was he constructing an elaborate symphony from individual melodic statements? Taking a long recording of one bird from the Eastern MacDonnell Ranges near Alice Springs, I set to exploring these questions. 🐦

I first began by notating his repertoire, finding this individual bird had eight melodic phrases. Some of them seemed to be the basis for frequent improvisation while others were never ornamented. It turned out that one phrase, what I came to think of as his 'home phrase', was repeated quite often, while others were less so. Curiously, one phrase was only sung a single time and not repeated.

However when I came to examining the sequencing, I could not identify any pattern at all. It seemed quite unpredictable which phrase the bird would jump to next. The only observation I could make was that he never sang the same phrase twice in succession. I had to concede I had no idea of what this bird was doing, if anything. Maybe there was no intelligence to his song, maybe he was just singing at random. And yet the precision of each song phrase suggested otherwise.

Disappointed, I put the project aside for some time, until I was next speaking with my colleague, the zoomusicologist, Hollis Taylor. Hollis is a violinist and researcher for whom Pied Butcherbirds have been both an enduring focus of enquiry and a source of creative muse. Over nearly two decades, she has travelled the country on extended field trips to record their songs, returning to compose instrumental music shaped by their melodies, phraseology, textural sonics and ornamentation. A resulting double CD of her music is entitled *Absolute Bird*,[18] and her insights and reflections on the species have resulted in a PhD thesis and the book; *Is Birdsong Music?*[19] If anyone could shed light on what butcherbirds are doing, it would be Hollis.

After describing my inconclusive results, she told me that she and her research colleagues had found the same lack of pattern in the sequencing of phrases. However, there did seem to be a pattern in the motifs contained *within* the phrases.

A motif can be thought of as a core component of a phrase;

a recognisable string of elements, but only part of a full phrase. Hollis suggested I go back to my data and focus on this aspect of their songs. Examining the eight phrases more closely, I was surprised to find that there were in fact four different motifs embedded within them. Remarkably, they were neatly distributed; each motif being found in two separate phrases. Another way of describing it would be to say that each motif, elaborated with additional elements, formed the core of two unique phrases.

Now I focused on only these four motifs, counting the two phrases made from them as variations of the one utterance. After re-analysing the sequencing, I sat back with a smile. Hollis was right, the pattern was clear and obvious.[20] Not totally predictable, but enough that it was undeniably intentional. She had cracked the code. Pied Butcherbirds do not sequence their songs randomly after all, but intelligently.

Hollis also pointed out that not only was there a sequencing pattern evident, but Pied Butcherbirds sing with a very specific degree of complexity. To appreciate the significance of what they are doing, we need to understand the way we appreciate art and music. Cognitively, aesthetics requires a balance between predictability and novelty. Take music for example: too little predictability in form and structure, and it may seem random, chaotic, atonal or rambling. As a listener, we might grapple with finding meaning, or give up feeling overwhelmed. Too little novelty, and we may easily get bored, as with a bland pop song. Not only does there need to be a balance, but each of us have our own preference and tolerance for predictability and novelty, our own aesthetic.

This understanding of human aesthetics suggests that Pied Butcherbirds also comprehend beauty. Each bird has its own repertoire of motifs. These provide the form and structure of its song – the predictability. The novelty, what we'd recognise in our arts as the creativity, is in how a bird plays with the motifs, elaborating them into several unique phrases, sequencing those and adding ornamentation. These are the sonic possibilities that represent each Pied Butcherbird's aesthetic.

There may also be, for the bird as with us, a sheer delight in sound. To our ear, butcherbird song is sonically pleasing. It may well be for the bird too, informing tone and melodic phrasing.

Overlaying all this is the geographic context of Pied Butcherbird songs. The range of the species extends over much of the continent, but regional populations display differing degrees of song complexity and size of repertoire. The ones I'm familiar with in eastern and central Australia are wonderful singers, often possessing around a dozen or so song phrases plus ornamental variations. However I recently spent time exploring the far northwest, near Exmouth, where ancient limestone hills are eroded into gorges near the coast. Here I recorded a bird whose repertoire was limited to only three phrases, with no common motifs and very little ornamentation. He sounded sonorous, but didn't play much with his song structure.

Mentioning this to Hollis, we wondered whether this was due to the aridity of the country and sparse population density. Spaced out along the range, those Pied Butcherbirds may not have interacted with each other very much. Without this contact, their stimulation – perhaps inspiration – to sing in more complex ways might have been diminished. Like a musician left to themselves without the challenge and invigoration of performing with others, their expression had become limited.

If so, then not only are Pied Butcherbirds displaying a sense of aesthetics, but the expression of individual birds is shaped by the community in which they perform. Collectively, Pied Butcherbirds could be thought of as affirming, maintaining and elaborating a singing culture. This suggests an intriguing possibility – that across the continent, these birds may be forming extended webs of community, sustained by an acoustic art, the richness of which is in direct response to the habitats in which they live.

I can't help comparing this with Aboriginal cultures whose stories, songs and dances grow from a relationship with place, threading the landscape into songlines. Perhaps Pied Butcherbirds may be thought of having their own songlines, criss-crossing the continent, essential to their being, and having been sung and maintained over vast timescales by generation upon generation of birds.

Superb Fairy-wrens

We've met a group of Superb Fairy-wrens on our walk, giving contact calls and 'happy family' vocalisations. Their actual song however is a bright, brisk, trilling reel, a brief shower of high frequency twitters.

One of the earliest recordings I made of a Superb Fairy-wren was of a male on a clear, spring morning. In an open patch of forest, he was illuminated by the sunlight, his iridescent blue and black plumage glistening. And he was singing; one of those rippling cascades of sound being uttered every few seconds. It was the height of the breeding season, and I recorded him for nearly ten minutes, which is an unusually long time for a fairy-wren to be singing continually.

Many years later, I was invited by a local naturalist group to give a talk on birdsong, and in preparation, revisited that fairy-wren recording. I thought it would make an interesting example of a familiar and much-loved local songbird.

Using spectrogram software, which renders sound into a visual representation, I analysed one of the songs and was struck by the complexity of the sonic structure.[21] What sounded to me as a pleasing jumble of twitters was incredibly acoustically detailed. I could see it comprised a dense series of whistled elements, delivered so quickly that they coalesced to my ear as an undifferentiated tumble of high frequency sound.

I paused to consider this. To deliver its song accurately, this bird had to be conscious of what it was singing. It couldn't be singing randomly, but shaping every detail of its song intentionally.

Realising that I couldn't comprehend what this bird was accomplishing in real time, I decided to try slowing down its song to a speed that my ear could follow. But by how much? It occurred to me that Fairy-wrens live in the wild up to around ten years – so an eighth of our lifespan. Turning it around; they live their lives eight times as fast as us. If I were to slow one of its three second phrases by a factor of eight, this would expand it out to half a minute and bring it down three octaves in pitch.

As I played the processed soundfile, I was aware that I was likely perceiving the fairy-wren's song in a similar way that it did. What had previously sounded like a chaos of twitters, was revealed as a series of pure, whistled tones, inflected up and down in discreet sweeps. The intonation and precision with which the bird did this was jaw dropping – what vocal control! The phrase began with a few tremulous, introductory notes, picking up pace to become a roller coaster of finely articulated trills, some higher and others lower, as though the bird was holding a two-sided conversation with itself. The more I listened, the more nuance I detected in the phrasing and pitching. It felt to me that what I'd uncovered was not the brief snatch of song lyric I'd expected, but a finely crafted poem.

I was now looking forward to playing this for my local audience, and began listening through the other song phrases on the recording. This led to more wonderment. Unexpectedly, I found that each phrase was different. What sounded to my ear like a succession of characteristic fairy-wren songs, were revealed as each entirely unique. Sonic snowflakes – there were no two alike. This bird was approaching every utterance anew, recasting his voice into novel songs each time. I thought of the ten minutes of vocalising I'd recorded of this one bird that morning – it represented not just a single poem, but an epic song cycle.

I'm not just saying this to wax lyrical. The amount of information contained in that fairy-wren's song was considerable. What this complexity of sonic story telling may represent to the fairy-wren I can only guess. What I did realise though, is that this bird's songs reflect a fundamental fact of the avian mind. It is quick.

Packed into a tiny skull, the neural connections of the compact avian brain are so closely wired that information is transferred at a rate that we literally can't comprehend. When you watch birds, with their alacrity and lightning-fast reflexes, you are witnessing their agile mental capacities in action. To creatures as keenly alive to the world as they are, large mammals like us must seem dismally slow-witted.

Slowing avian vocalisations to our pace of comprehension brings me to reflect on time itself, which we quite naturally perceive by our own human measure. Yet it is different for other creatures. Each must experience time according to its own rate of cognition. For a bird, their sense of it will be rich in granular detail. Perching for a minute may seem like an extended period of stillness for them, and a single day pass like one of our weeks.

In light of this, I find it wonderful to think of other creatures and realise that each experiences the temporal dimension differently. No longer do I assume the passage of time as being shared in common, with all life traveling at the same pace. Instead, like multiple lanes of traffic moving at different speeds, I sense an ecosystem of creatures each perceiving time in their own unique way. Observing dragonflies zipping over a pond, I see them changing direction, speed and altitude faster than my eye can follow.[22] Birds, I imagine, take it as quite acceptable that most things move sluggishly enough for their quick reflexes to react. A flight of lorikeets, darting at breakneck speed through the trees like so many green torpedoes, may perceive themselves leisurely weaving their way among the foliage. And a fairy-wren, basking in the long hours of morning sunshine, has ample time to embark on an epic song cycle.

Listening allows us awareness into these parallel universes of time and consciousness. Hares and tortoises, birds and humans, we all share this planet. Yet it is sobering to consider that, as Dr Who reflects, time is ... "Complicated. Wibbly wobbly, timey wimey ... stuff ..."

Interlude

The Valley of the Winds

Wind is the nemesis of nature sound recordists, and I had just about had enough of it. In the Australian outback, where flat country and extremes of temperature can result in blustery conditions at any time of day, it can be particularly frustrating. But never more so than around dawn, when the air should be calm.

We were recording at Uluru, that iconic landform situated at the heart of the Australian continent. This was a few years after our decision made under the storm-lashed tall trees of Waratah Flat, and we were by now working on producing a collection of nature recordings from environments around the country. Having gained permission from Aboriginal custodians and parks authorities to record, we were allowed to enter the park during the curfew hours before dawn. One morning I decided to try recording at the similarly distinctive rock formations of Kata Tjuta, an hour's drive to the west, requiring a start in the small hours.

As we set off, the conditions were encouragingly still. By the time we arrived at Kata Tjuta's carpark however, a wind had arisen from nowhere and was howling. This was no breeze. The shrubs were being whipped around, and dust thrown up in gritty swirls. Sarah and I discussed whether it was even worthwhile me record- ing here this morning. I decided I'd give it a go, and that she would return to pick me up after photographing elsewhere.

Kata Tjuta is, like Uluru, a remarkable location. Domes of bare, ochre-red sandstone emerge from the surrounding sandhills, grouped together in rounded forms bisected by narrow, vertically- walled gorges. As I walked the rough path across the open country toward these formations, their silhouettes sat on the eastern skyline, looming larger as I approached. All the while, the wind didn't let up. Following the track as it wound among the nearest hills, I knew recording would be pointless. Nevertheless, attempting to find a sheltered place, I stood near a twisted bloodwood tree, recording the wind thrashing its branches with only a few plucky honeyeaters chipping briefly to signify a dawn chorus.

As I was here now, I figured I may as well go exploring. Wandering further among the rock domes, I entered a narrow valley between sheer, weather-worn cliffs. In the deep shadows, the only indication of sunrise was the very top of the walls opposite turning a vivid crimson with the first rays of sunlight. I watched as the flaming line inched down the rockwall. Red rock against blue sky – a quintessentially Australian sight, I've never witnessed anything quite like it elsewhere.

As impressive as the view was, I was nevertheless disappointed at not being able to get any useful recordings. Finally acknowledging the morning would be fruitless, I chose a somewhat sheltered location and sat down.

I found myself remembering an occasion at Mutawintji, when the local Aboriginal community had arrived for a weekend meeting, and invited Sarah and I to join them for the evening meal. It was the first occasion we'd eaten barbecued kangaroo tail, and the last time we ate boiled emu. I smiled, recalling the traditional recipe for cooking emu: cut the meat into cubes, put it in a pot, season with salt and curry powder perhaps, cover with water and throw in a stone. Cook for several hours. When the stone is soft, throw out the emu and eat the stone. It's an old Aussie joke, with more than a gristle of truth.

In the course of chatting with the Mutawintji mob over dinner, Sarah and I mentioned that we'd walked to a striking landform known locally as Split Rocks. Boulders the size of multi-storey buildings seemed calved off the edge of a rocky ridge, as though set there balanced on end. It was an unusual location, but eerie, the kind of place you feel your skin prickle. The story came out that it was a site of great significance for Barkindji people, but the last time they'd taken a bus load of kids out there for a ceremony in the evening, a huge wind had started. Dust was kicked up, getting in their food, and the wind was whipping everything about. The kids were all painted up and ready, but the wind was still howling. One of the elders eventually spoke up: "Spirits didn't want us here this time". They abandoned the ceremony. As soon as they'd decided this, the wind dropped to a dead calm. They packed up and departed.

As I sat there, remembering this story, the wind was still gusting. Kata Tjuta is a deeply sacred place for local Anangu People. I began wondering whether the wind was trying to tell me something. I began to question my motivation in being there. By making recordings, wasn't I exploiting the land the way white fellas so often do? Wasn't I taking from nature to produce a commercial product? Maybe the spirits of the land, if they existed as the Aboriginal people understood them, didn't want me there. Perhaps the wind was giving me a message that I wasn't welcome.

If one can reason with spirits, I tried to be honest about my motives. If I was pursuing financial return, there were surely easier and far more lucrative ways of doing so. I knew that, in my heart of hearts, what drew me to this work was my desire to share the beauty of nature with listeners, some of whom would never get to experience these wild places for themselves. I thought back to a conversation we'd had with the nature photographer and publisher Steve Parish, who'd advised us; "People won't protect something they don't first love". I was hoping our recordings would inspire a love of wild places and the creatures that lived there. Being out here, capturing authentic recordings, was necessary and integral to that purpose.

A part of me was thinking that this was a silly and vaguely superstitious line of thought to be entertaining. Nevertheless, I pursued it further. If my intentions were genuine, then why this wind? Maybe I was simply being asked to 'come back another day'? That seemed rather trivial of the universe.

Then it occurred to me to flip it around. Maybe, instead of not being here, the tricky conditions were an invitation to be *more* here. All that morning I'd been telling myself there was nothing worthwhile to record. Now I wondered if I'd been missing something all along.

I began listening more closely, really giving it my full attention. I was now in a more sheltered spot, and at first, didn't notice much in particular, not even any birdsong. I soon become aware of something though; a pervasive roar. It was soft, constant and seemingly distant. Rather than being all around me, I got the impression it was coming from above. Then I realised what it was; the sound of wind buffeting against the rock walls. Over millennia, similar winds had eroded the cliffs, pockmarking them with indentations, ridges, crags, caves and overhangs, all of which tore the wind into vortices and turbulences. It sounded like distant surf, ebbing and swelling imperceptibly as it surged from one cliff face to another.

I then heard, contrary to my earlier perception, that there was some birdsong. However it was mostly so far away that it arrived to me as disassociated echoes drifting down the narrow canyons. To my surprise, here was a whole soundscape I'd been unaware of. I switched on my microphones, and for the next hour delighted in the smallest of acoustic details. The song of a Grey Shrike-thrush drifted from some hidden place. A flight of Painted Finches passed by, their faint calls barely audible. A Common Bronzewing pigeon began a soft booming not far off, its voice muffled and smudged by the topography. A Nankeen Kestrel revealed itself as having a nest in a high crevice, flying in with excited chittering which reverberated off the rocks.

I was so absorbed, that I ended up recording far longer than was my habit at the time. What began with feelings of futility, transformed into a very satisfying and rewarding occasion. The recording, albeit in edited form, ended up as a track on our *Uluru* album, appropriately named after the location; The Valley of the Winds. 🐦

Even at the time, I was aware of a different quality of listening that morning. Previously I'd been thinking in terms of species, with the aim of obtaining clear recordings to represent the biodiversity of an environment. This seemed an appropriate way to proceed, both in working with microphones and listening. As the wind whipped across the landscape that day, I had concluded that turning on the mics, or even listening particularly, would be a waste of time.

In the process of assuming this, I tuned out from what was actually happening around me. Even when I found a relatively sheltered place where the microphones would not be buffeted, it was my state of mind that prevented me from thinking there was anything worthy of interest. I recall the thought that pulled me up; in nature, there is always something to hear, always, if you listen for it. Of course, I appreciate there are times when it really is quiet. The depths of night in temperate latitudes can be one such time, and I've visited a few truly barren locations. Even then, I find stillness has a presence to it.

But the significance of that morning was not really about sound, it was my lack of attention. I was surprised at how easily I lapsed into a lazy awareness, and how surprising it felt to reclaim my attention.

When I listened back to the recording I'd made, there was a further discovery. I found that I'd captured something I hadn't expected. In those distant and echoing shrike-thrush calls, the fragile voices of Painted Finches against the wind, the muted bronzewing and the kestrel's cries reverberating off the rock walls, I'd documented something unforseen – the landscape itself.

The wind sheering off Kata Tjuta's rockwalls was one part of it, that sense of the elements characteristic of the place. But what surprised me was hearing the voices of birds shaped by the landforms among which they lived. I could hear the vastness of the landscape, the hard, reverberant quality of the rocks, and the sparse, distributed populations of birds. The environment in which these species lived was vividly audible.

For the first time I was hearing the world as integrated, rather than a collection of discreet voices or species. I was listening to the physical landscape, the weather conditions, and the acoustic context of the place. I was hearing rocks and sky as much as feathers and song.

This holistic quality of listening may seem a contradiction to the species and repertoire analysis I have regaled you with during our walk along our back track. Of course, they are complementary and both valuable. Perhaps we could say that in listening to species, we're hearing the narrative threads, while the soundscape allows us to take in the whole story, hear how those narratives entwine, and put it all into a context of time and place.

I hope I'm giving you a sense of just how profound this shift in perspective was for me. I discovered that it was my expectations that inhibited authentic and deep listening. By 'authentic', I mean simply hearing what was around me without preconceptions. Letting go of those judgements allowed me not only to notice more acutely, but to tune my awareness to the nuances of my surroundings.

Pursuing this way of listening not only influences one's state of being for the better, it also supports a different perception of the natural world, one in which relationships and interconnections become more apparent. I've mentioned previously how we conceptually separate the world into objects. This was the first occasion that attentive listening allowed me to 'put the world back together'.

Before we move on, I need to say a few more words about the process that precipitated this shift.

That morning at Kata Tjuta, I reflected on what I understood of Aboriginal people's relationship with place, and how connected they are to the land. No doubt my knowledge of Indigenous thought was, and remains, superficial. However in thinking of nature as a living presence, I began considering myself in relation to it. This took the form of a conversation – a searching and rather personal one. What was my intention? Was I acting with integrity? What did the place have to say to me? I reflected on my purpose and whether it was welcome, and sought some response by paying attention.

If I hadn't been pursuing these lines of thought in discussion with the land – a very non-Western way of conceiving one's relationship to nature – I wouldn't have come to the awareness and subsequent realisations that I did. In my culture's way of describing such things, I'd call this a communion.

I could have walked out of Kata Tjuta that morning with a grumpy, grey cloud over my head and no recordings. Instead, I believe I had my first glimpse of what Aboriginal people talk about. I discovered an awareness of nature that is transpersonal. Nature talked to me, told me things, and I heard a new way of being.

Chapter 5

Voices of the Land

A Northern Puzzle

We'd been travelling across Europe for two months recording the spring birdsong, and I was perplexed. Not that Sarah and I hadn't had a rewarding trip – far from it. Arriving in Istanbul in late March, I'd begun by recording dawn birdsong in the peaceful garden of the Blue Mosque, with the call to prayer drifting across a waking city. 🐦 We'd then set off on a grand circumnavigation of Turkey.

On the World War I battlefields of Gallipoli, the songs of Blackbirds, Goldfinches and Sardinian Warblers were a peaceful contrast to the horrors of warfare a hundred years previously. 🐦 Going back further in history, we witnessed how nature had also reclaimed sites from Greek and Roman times. At the ruined citadel of Termessos, Jays hopped among fallen masonry 🐦 and Rüppell's Warblers called out across the precipitous mountainsides which had once made the city impregnable. 🐦

We recorded Nightingales singing on the banks of the Euphrates River, and the songs of larks drifting across fields near the remains of Hattusa, the capital of the Hittite empire around 1500 B.C. 🐦 In the Taurus Mountains, birdsong mingled with the clinking of traditional sheepbells, 🐦 and in the far northeast, the soft cooing and slapping wings of Feral Pigeons, a common species I'd normally overlook, reverberated within the roofless shell of the thousand-year-old Ösk Monastery. 🐦

As we travelled, I was fascinated by hearing nature within a human context. This was a novel perspective for me. In Australia, the land has a cultural history that is far more ancient, however that Indigenous story is not apparent to the uninitiated. There are no monuments, decaying temples or crumbling walls to echo the birdsong. Instead, as I'd found at Kata Tjuta, the land speaks directly.

Despite finding nature and place evocative in Turkey, I was unexpectedly struggling to find an aesthetic in what I was hearing. We were encountering rich, spring soundscapes, yet something was eluding me. European birdsong was unlike anything I was familiar with in Australia.

It was more than simply different species; they were doing something different. Frequently I'd observe a songbird perched high in a tree, often at the very top, pouring forth a complex repertoire. The virtuosity of song, and this habit of delivering from the highest perch, was something I'd never encountered before. In Australia, I was familiar with birds singing in simpler ways that created patterns of sound in the landscape. But in Turkey, these pleasing patterns were absent, replaced by what seemed to be a dense and undifferentiated twittering.

From Turkey we followed the advancing spring north to the forests and upland taiga of Sweden. Finally, arriving in the UK by midsummer, we visited our friends Jane and Geoff Sample in Northumberland. Geoff is an experienced and sensitive nature sound recordist, and generously, he took me to some of his favourite local haunts. Heading out early one morning, we arrived at the edge of woodlands overlooking a broad and shallow dale. Assessing the lie of the land, I decided on placing my microphones at a spot I anticipated would result in a panoramic recording.

I was curious how Geoff would capture the same landscape – this was his native place, after all. I returned to ask him where he'd put his equipment. "Over there", he said, pointing up a grassy slope toward a solitary bush on the skyline, his tripod and microphones just visible at its base. He'd run a long cable some two hundred metres back to where we stood, and was waiting to switch on his recorder. Now I was more puzzled than ever, and trying not to sound too naive, asked his intention in doing so. "Well …" he began in his warm, north country accent, "there is a Linnet which uses that bush as a songperch, and every half an hour or so he might come in there and sing for a minute or two. If I'm lucky I should get a nice recording."

Geoff indeed had an entirely different ear on his native birdlife than I did. Back in his studio, he played me one of his favourite recordings, of a single Thrush Nightingale singing at 2 a.m. in a Finnish woodland, its complex phrases emerging from the shadows of the twilight.

Now I could put my finger on what had been baffling me. I am used to birdsong in Australia being given collectively, in chorus. In Europe it was solo and virtuosic.

We discussed this at some length, as Geoff spoke of the obvious; songbirds in northern Europe are adapted to a short, summer breeding season. Often they migrate from their southern wintering grounds as far away as the Sahara or East Africa, dispersing over Europe and arriving in the north around April. They immediately establish homeranges, the males singing vociferously both to delineate their patch and attract a mate. They have a brief window of opportunity when food resources are abundant, during which they must raise a clutch of half a dozen or more chicks, and depart again before the autumn cools.

In adapting to such a brief but plentiful season, they have little need of developing or maintaining elaborate social networks. Nor do they even have to nurture an enduring pair bond with their mate, as it may be dissolved once the current season ends and they migrate southward. They just need to get on with it quickly so their chicks can be independent by the time the weather turns. For northern birds, getting on with it often involves showmanship, and what better stage than the highest songperch in the landscape?

It was not just different species that I had been recording across Europe, but a whole different way of living. What I was hearing were the solutions of songbirds to the challenges and opportunities of northern geography and a highly seasonal food supply.

I was beginning to realise that what ultimately shapes birdsong is the land and climate. In other words, the sounds we hear in wild places are shaped by factors other than the vocalising creatures themselves.

Antipodean Aesthetics

The Little Desert is a swathe of sandy country in Australia's southeast. Created over millennia by prevailing winds pushing coastal sands inland into a jumble of low dunes, the region is a mix of open woodlands, broombush, mallee and heathlands. Like many areas of low rainfall, it is floristically interesting, and this diversity attracts a range of birdlife.

I've returned to the region several times over the years, partly lured back by my first visit which yielded one of my favourite recordings. In the first light of day, among dense, scrubby heathlands that stretched as far as I could make out, I chanced upon a dawn chorus of Tawny-crowned Honeyeaters. Individually, each tawny-crown has a pleasant song; a series of delicate, whistled notes with a winsome quality. As delightful as this song is, when a group are heard singing together, as they do at dawn, they're nothing short of magical. As I recorded that morning, all I could hear across the landscape were the songs of tawny-crowns woven into an intricate harmony. When people hear this recording, a frequent response is that it sounds more like synthesised music – electronic, not organic at all. But birdsong indeed it is.

Making this recording early in my vocation consolidated an aesthetic that had been developing ever since hearing those Spiny-cheeked Honeyeaters at Mutawintji. If I could attempt to put my finger on it, I'd say it was a simple repertoire sung by multiple birds in chorus, creating pleasing sonic patterns.

I'd find examples of this patterning wherever I went across Australia. In the outback, it could be the sweet, down-slurred whistles of Chestnut-rumped Thornbills. In the rainforests of far north Queensland, it was the loud, chopping, percussive vocalising of appropriately named Chowchillas. In our south-eastern forests, the constant 'pings' of Bell Miners create a unique sonic effect, and almost everywhere across the continent, various species of honeyeaters give our dawn choruses a sweet coherence.

In those early years, as I recorded widely across my homeland, it did not occur to me to question why our birds were creating the sonic patterns that I so enjoyed focusing my microphones on. I took it for granted that these constellations of song were what birds do. Only after returning from Europe did I appreciate the degree to which Australian birdsong is different. What began for me as an aesthetic awareness now prompted a sequence of questions. Why do our songbirds sound the way they do? Why do we not have as many soloist, 'song-perching' species with virtuosic repertoires?

I say this while being aware that it's not that simple. You will hear some skilful solo recitals from Australian songbirds. The Rufous Whistlers of our local bushland are effusive and accomplished singers in the warmer months. 🐦 Across our subtropical regions, Brown Honeyeaters have such a wide-ranging vocabulary as to prompt a colleague from Britain, Jennifer Beasley, to dub them 'Australia's Nightingales'. 🐦 If one listens attentively, one will occasionally hear smaller species also giving complex songs, often including tidbits of very adept mimicry – thornbills, hylacolas, Silvereyes, Mistletoebirds and Horsfield's Bushlarks come to mind. 🐦 And of course we can't mention mimicry without acknowledging lyrebirds as creating one of the world's most outstanding solo vocal displays.

Nevertheless, with these exceptions perhaps, antipodean birdsong would probably seem unimpressive to northern listeners. Australia has few spectacular show-offs, especially ones that pick the very topmost of a tree for extended sessions of grandstanding. This is not to suggest we're a continent of avian introverts though. Indeed, I find the singing behaviours of my native land to be among the most fascinating in the world.

Co-operative Lifestyles

Reflecting on my discussions with Geoff, I came to understand that Australian birdsong sounds different because our birds live differently. This often comes down to one factor: co-operation. Many Australian birds have developed sophisticated ways of living together, and as these strategies require co-ordination of varying kinds, they are facilitated by the ways they communicate.

We've heard examples of this during our listening walk; contact calls, long-term bonding songs, intimate vocalisations, family and social chatter – these all facilitate the more socially-based lives that many of our birds lead. Once I'd made this connection, my aesthetic response to Australian birdsong gained a behavioural context. Chestnut-rumped Thornbills glean through desert foliage in animated flocks, creating between them a delicate swirl of activity. Chowchillas live in family groups and define their homeranges with bouts of calling in which individuals take turns, filling the air with a continual and raucous chorus for extended periods. Bell Miners form colonies numbering dozens of birds, occupying a patch of forest in which their collective, incessant calling throughout the day acts to deter other birds from entering. Honeyeaters of many species are resident in loose communities, or nomadic in small parties, roosting and singing together at dawn.

While they may live communally or travel in groups, these birds still establish regular pair bonds for breeding, a male and female incubating and hatching a clutch of chicks, feeding and caring for them until they fledge and can fend for themselves. However a significant number of Australian species specialise in another reproductive strategy, which generates its own range of vocal behaviours – co-operative breeding. In this co-operative model, a whole extended family gets involved. In addition to the parents, a number of helpers – siblings, young adults or even non-kin adults – all assist in rearing the brood. This form of breeding is often accompanied by vocal performances that unite parents and helpers in a co-ordinated effort.

Laughing Kookaburras

A fine example of this team building in song is that most widely recognised of Australian bird sounds – the rollicking calls of Laughing Kookaburras. These are given in chorus by several birds at once, often perched close to each other, heads thrown high, tails cocked, their throats pulsing with hoots and rawks and cackles.

One bird will begin with quiet chuckles, which are picked up by the others. As everyone joins in, the chorus builds to a cacophony of shouted *koo, koo, KOO, KOO, KOO, KA, KA, KAKAKAKA* ... After a raucous crescendo, they'll slow and wind down to some final chortles.

It is both a spirited and synchronised performance. This chorusing group will be a single, family unit; an adult couple plus younger helpers of both sexes drawn from previous breeding efforts. While it may not seem that the racket they create is conducive to co-operation, it is sonically complex, and as we've recognised, this quality of sound is often a good vehicle for social communications.

Families of kookaburras are frequently among the first birds to sing before the dawn, signalling the beginning of a new day. They're also the last to roost, gathering together for a dusk chorus in the twilight before they turn in. Calling may also be initiated at any time during the day. If the family is dispersed, other members hasten to regroup, flying in to join the party. The volume of sound they create together is ideally suited to carrying across their extensive homeranges, and often elicits answering responses from neighbouring families.

White-winged Choughs

Australia has become recognised as one of the co-operative breeding capitals of the bird world. Pointing to just how common it is, we have several co-operatively breeding species that live in the bushland around our home. Laughing Kookaburras are one, as are our Varied Sittellas[23]. Another are White-browed Babblers, who live in social mobs of up to a dozen birds, roost together in shaggy, domed, stick nests, and are constantly vocal; chattering, scolding, wheezing and whistling to each other as they forage on the ground, or chase each other around low shrubs. The families of Australian Magpies who sing together with seemingly telepathic accord are also social breeders. Plus there are two others we've already met; White-winged Choughs and Superb Fairy-wrens.

We've seen that choughs live socially, rarely being found out of each other's company. This life-long mode of living is integral to their breeding strategy. A tribe of a dozen or so choughs will usually consist of a breeding adult pair, plus offspring, either of their own or adopted from a neighbouring group. They will build a robust, mud nest on a high horizontal branch, taking turns incubating and then raising young ones collectively. Younger birds will stay with their family for several years, helping out with junior siblings.

Any time spent with these birds reveals them as playful and social. Adults (recognisable by red irises) and younger ones (with brown eyes) both get involved in collective games. Sometimes these are spontaneous, and at others quite formal behaviours. One is known as 'goggling'. Adults gather, often around young ones, and engage in a display that involves wagging fanned tails up and down and half extending their wings. They'll simultaneously draw back facial feathers to accentuate blood-engorged conjunctiva around those vivid red eyes, all the while giving soft, mewing calls. It's quite a dramatic show.

They also delight in playing with objects. Puzzled, I've followed a tapping sound in the bush to find a chough with an old, dried up landsnail shell in its beak, hitting it repeatedly against a stone. They seem to do this for no practical purpose, simply the fun of it. A neighbour of ours has choughs that favour the white quartz pebbles of her driveway, and she no longer parks her car near their regular place of stone tapping. Her neighbour's birds have taken to spinning around the wooden pegs on their clothes line. (This is preferable to a Sulphur-crested Cockatoo that learned to methodically unpeg an entire line of washing, watching each garment drop to the ground.)

Another amusing quality of choughs is that they are goofy. Alighting, they will wag their tails up and down as though trying to recover their balance. They're the only bird I've seen fly up to a branch – and miss. Or, making their way along a branch, fall off. The squawks of indignation that accompany these missteps are as though accusing a tree of unreliability.

My most endearing chough moment though was quite remarkable, and something I've only observed on two occasions. If you can picture half a dozen birds, each lying on their backs on the ground, snuggled up side by side as if in bed together, with their feet raised in the air, and passing a small stick back and forth among them. They continued this for a minute or so, until one seemingly got bored and discarded the stick. Then they all hopped up, game over, shook themselves off and resumed their usual activities.

All this speaks of a bird that has character. Their curiosity and play has a purpose; it is a way of both learning and bonding. Choughs need to work together, and intelligence (goofiness notwithstanding) is part of the successful formula. Their constant communications are essential to this way of life; a combination of practical repertoire such as alert and bonding calls, and those of a more individual and expressive nature. It is relatively easy to tell the difference. Their formal calls, notably their 'yellow alert' whistles, are shared and specific to circumstances. Their individual vocalisations are more variable in character, affording them a wide expressive palate.

Interestingly, while kookaburras often begin calling boisterously at first light, choughs don't much at all. If anything, they may be heard saying good morning to each other with a short fuss of whistles, but that's it. This is in contrast to the remainder of the day when they rarely shut up.

Superb Fairy-wrens

Superb Fairy-wrens approach co-operative breeding differently, and one can hear this in their vocalisations.

In the stillness of the predawn, well before the dawn chorus begins, one will frequently hear the trills of Fairy-wrens. This is noticeable across the country, wherever these delightful little birds live. At Mutawintji I was recording the White-winged species at around 3 a.m., with the stars still bright overhead. At home, our Superb Fairy-wrens roost in a dense and sprawling shrub near our bedroom window, and to get a good night's sleep in the breeding season we have to habituate ourselves to sleep through their wee-hour festivities. 🐦

Fairy-wrens live in extended family groups, comprising one or maybe two males in their spectacular iridescent plumage, plus a number of sombre brown females and sub-adults. They all contribute to rearing the young, but the parenting itself is the ambiguous aspect. Fairy-wrens have been discovered to be notoriously promiscuous, with very few of a female's chicks being sired by the resident male. Instead, she flits off in the predawn hour to mate with a neighbouring male. How does she know where he is in the darkness? He's singing out for her, of course.

It would be easy to make interpretations around cuckolding and infidelity, but for fairy-wrens this is a necessary aspect of their breeding strategy. The female is simply accessing genetic material from further afield than her own family group, genetic diversity and resilience being the benefits. This behaviour may confront our nuclear family image of fairy-wrens, but it is an adaptation to such a tightly social form of breeding.

Once eggs are laid, the remainder of the incubation and rearing process is very much a family affair, with the extended group pitching in to assist. Here, co-ordination counts, and once again, it is vocally apparent. To hang around fairy-wrens is to see them constantly in motion, flitting here and there, darting from shrub to shrub, investigating dense vegetation and pecking at anything interesting. And much of the time, they'll be vocalising. We've already heard some of these sounds; delicate, intimate, I'm tempted to say affectionate. Watching them make these private sounds, one can sense their tight family connection.

Before long though, you'll see and hear them do something that perplexed early ornithologists greatly. A male will lengthen his neck slightly and begin singing, that splintered cascade of sound you're now familiar with, and which his whole body vibrates in producing. As soon as he starts, from nearby comes a supportive trill. This is from the female, her brown plumage similarly puffed in exertion. 🐦 While this male-female tag team effort is heard commonly, one may equally hear a female singing alone. That females would sing – and do so as strongly, often, and as competently as males – may have confused early European scientific expectations (as European birdsong confused mine), yet considering fairy-wrens' familial bonding, it seems appropriate that their singing would be shared.

The trilling of fairy-wrens may have several functions, but a primary one is to define their homerange. Around our house, there are at least two groups of fairy-wrens in the garden. We often hear them singing back and forth, one group to another, males and females swelling the sound between them in affirmation of their respective living spaces. In the middle, they seem to have agreed to share the bird bath.

Alternate Lifestyles – Emus

Female vocalising is just another way Australian birds tell us they live differently. Emus take this a step further.

In the stillness of the desert night in inland Australia, the calls of Emus carry for great distances over the plains. They sound like someone thumping a bass drum far off, a deep booming that can be almost subliminal. Unlike any decent drummer though, they have a poor sense of rhythm, bouts of calling being given staccato fashion, in stuttering sequences. It is a sound that is characteristic of the outback, and one I find familiar and reassuring.

When you hear Emus like this, you'll be hearing the female. She is the one who gives those booming notes. She calls, attracts, seduces and eventually mates with the male. After eggs are laid in a shallow nest on the ground, she often wanders off, and has little further involvement, leaving the male to incubate, hatch and parent alone.

He does this well, and can be a fierce protector of his brood, as we were reminded one outback morning. Sarah and I, ambling around the scrub, were approached by a group of curious sub-adult Emus. It was thrilling to have these wild birds so close, yet we cautiously kept to some low shrubs so as not to spook them. Distracted as we were, I heard a thin, quavering whistle from behind me, and turning, was surprised to find eight Emu chicks curiously eyeing us from nearby. They are adorable when so young, being downy white with several black speed stripes down their sides. I crouched down for a good camera angle, at which they milled around photogenically. When I looked up from the viewfinder though, it was to find their attendant parent approaching quickly – and he was not happy. Neck feathers extended in a big ruff, he was striding purposefully toward us, head and body stretched high. It was one of those moments – we've had a few of them – when one suddenly realises you're in danger. Emus can be intimidating and their clawed feet dangerous. Fortunately, the chicks took the hint and quickly moved away, dad turning and shepherding them off.

So for Emus it is the female who is the vocal one, while the males are relatively silent. Emus are not unique in this, with female song and seduction also initiating the breeding of other species, including Australia's button-quail, Painted-snipe and the idiosyncratic Plains Wanderer. Like the Emu, these birds are all descended from ancient lineages, bringing me to wonder whether this unusual behaviour is in some way ancestral in birds, and possibly even inherited from their theropod dinosaur forebears. Did the females of those fearsome reptiles also take the romantic lead while leaving parenting commitments to the males?

The Contrast of Hemispheres

The Australian naturalist Tim Low sums up the diversity of our bird behaviours and lifestyles with cheeky humour. In his public presentations, Tim shows a slide with an outline map of the world. Across the continents of North America, Europe and northern Asia is written the word 'boring'. The southern landmasses – southeast Asia, and fragments of the ancient Gondwanan supercontinent comprising Australia, Africa and South America – are labelled 'exciting'.

In his writing, Tim has speculated on why Australian birds have found such a range of interesting ways to live.[24] The reasons are no doubt complex and many, but one aspect centres on Australia's unique environmental circumstances. Being an ancient landscape, the continent's soils are generally leached, poor in nutrients, and patchy. Australia's plants have adapted to them, but in the process become specialised and restricted in distribution. While northern hemisphere plants, living on glacially homogenised soils, were able to spread in response to climate swings, especially during the Pleistocene, by producing nutritious seeds dispersed by animals and birds, Australia's plants have had to evolve in situ. A deficiency of the soil nutrients required for seeds, but an abundance of sunshine to create sugars, has led our flora to invest in nectar and pollen, which is easily transferred to favourable sites.

To the flowers of wattles, grevilleas, banksias, bottlebrushes and eucalypts are attracted a fauna of native species – not just birds but insects and small mammals – who are at least partly reliant on nectar, and through whose attentions plants are able to disseminate their pollen widely. Our numerous and diverse honeyeater species, plus conspicuously noisy lorikeets, are an outcome of these associations.

To soil deficiencies, can be added climatic ones. Over geologic time, the Australian continent has drifted north into latitudes poor in rainfall, resulting in much of the inland becoming arid, semi-desert. Precipitation in these regions is erratic and unpredictable, and when it does come, it often arrives in abundance. Australia's plants and animals have adapted to this too, with significant rainfall events resulting in a flowering of the desert and stimulating a fever of opportunistic breeding.

This may paint a picture of the Australian continent as a harsh place to live, driving the development of intelligence and co-operation as a means of surviving in challenging circumstances.

But as Tim suggests, and I tend to agree, another interpretation is more likely. Australia is a very diverse landscape, especially when compared to those northern regions of Eurasia and America. It encompasses habitats ranging from ancient temperate and tropical rainforests to deserts with a surprising diversity of vegetation. There are woodlands, heath communities, forests and grasslands of great complexity, plus wetlands which range from as reliable as the tropical wet season to ephemeral systems that erupt into life after rains. Across the continent, habitats are characterised by an evergreen vegetation providing a year-round insect supply, while the climate lacks harsh winters.

These factors combine to make Australia a benign place to live, offering many opportunities. Birds are known to have been present in southern lands for a very long period, giving them ample time to refine intelligence as they've adapted to relatively stable circumstances. Continuity has allowed resident lifestyles to emerge, with life-long pair bonding, social dependence and co-operative breeding being among its many expressions. Even species such as our inland honeyeaters that have adapted to moving around semi-arid regions, have brought collective living to a more nomadic lifestyle.

From this perspective, co-operation and intelligence can be understood as having emerged due to stability, rather than challenge. Australia is by no means unique in this, with many of the behaviours characteristic of the southern continent being found elsewhere. In the Himalayas of Nepal, I was delighted by the antics of parties of a dozen or so White-throated Laughingthrushes, foraging together on the forest floor and keeping up a constant chatter that reminded me of our choughs and babblers at home. 🐦 In Thailand, several species of Laughingthrushes live similarly, joining in making a cackling ruckus which resonates through the rainforest. 🐦 In East Africa, White-browed Robin-Chats, Spotted Morning-Thrushes, Ground Hornbills and Red-and-Yellow Barbets are among a suite of species that match our butcherbirds and magpies in bringing the vocal duet to a sublime art. 🐦

So in a global context, the soloistic showmanship of northern songbirds is unusual. Virtuosity and extended repertoires can be interpreted as specific responses to the brief abundance and bitter winters of the north. Extreme seasonality acts to limit behavioural diversity, negating the benefits of social bonding and replacing them with the obligation of long-distance migration. In more climatically homogenous bioregions, birds have found, as Tim Low puts it, more 'exciting' ways of living together.

This is not to contrast co-operative with 'competitive' ways of living. While many northern songbirds may appear to be engaging in contests of song, I believe this is a misinterpretation. Instead, they have adapted, as all creatures must, to the particular combination of opportunities and challenges they face, the solutions to which do not preference or require the development of co-operative behaviours.

There is much yet to be learned. For instance, we don't know whether co-operative breeding is an ancestral behaviour for birds, or a more recent set of adaptations. Considering how globally widespread these co-operative lifestyles are, it seems more likely they are ancestral. If so, then it suggests co-operation has been lost and replaced with ephemeral and more circumscribed relationships among many northern temperate species, as they've adapted to migratory lifestyles and the requirements of rapid breeding. In the process, social and long-term singing behaviours, such as duetting and female song, have also been lost.

Whatever the scenario though, I hear the birdsong of my native land as both fascinating and aesthetically engaging. Although not as florid as some northern species, I still marvel at the solo performances of Rufous Whistlers, lyrebirds and thornbills. I delight in the sonic patterns created by communities of birds singing in concert and find endless entertainment in the vocal interactions of more social species.

All these communicative patterns go to make up the overall soundscape, and have in turn been shaped by the habitats and climatic circumstances in which these birds live. While creatures fill the landscape with sound, it is the environment that has ultimately shaped their behaviours, and hence their sounds. As I instinctively felt when first hearing those Spiny-cheeked Honeyeaters at Mutawintji, the land itself is indeed audible. It sings through the creatures that live there.

A Foreign Idiom

When speaking in such generalisations, it is reassuring to occasionally come across modest confirmations. Recently, Sarah and I took a few days bush camping on the Glenelg River in western Victoria. One morning I noticed an animated stream of rippling notes and trills that were quite a contrast to our usual bush birdsong. Tracking the sound, I found a small bird singing its heart out from the very topmost branch of a tall eucalypt. It wasn't until I got my binoculars on it that my suspicions were confirmed. It was an introduced European Goldfinch, bringing its inheritance of elaborate song and preference for high perches to a new home in the antipodean bushland.

Islands of Wonder

The Monarchs of Tetepare

Sarah and I took our first steps onto the coral sands of Tetepare, a remote island in the western Pacific. We were glad to have arrived safely. After three days of sea journeys to get here, the final stretch over open ocean in a tiny boat had been a white-knuckle ride. Just as we approached the island, broaching the high waves on the island's outer reef, the outboard engine had cut out. As we wallowed perilously in the swell, our local boatmen, native to these waters and seemingly unperturbed, chatted leisurely about how to get fuel through to the cantankerous motor.

North of Australia, the Solomons are a string of tropical islands extending eastwards from New Guinea out into the Pacific. Of the many islands in the archipelago, the 40-kilometre length of Tetepare Island was one of few conservation reserves, and the only island altogether free of the rapacious logging that had degraded others and so adversely affected many Indigenous communities. As the pace of logging had intensified, the local Tetepare islanders, supported by two Australian ecologists, Katherine Moseby and John Read, set about protecting their island. The plan focused on establishing an eco-lodge and research base. After many setbacks and having come close to failure, the project has finally been a success.[25]

Now ashore, we were welcomed by local villagers and shown to our hut, a pole and grass thatch dwelling made in the traditional manner entirely without nails. After settling in, we met the man who had both constructed the building, and was to be our local guide. Twomey was quiet and thoughtful, with an intimate knowledge of the island and its wildlife. To have his support and guidance in sound recording his home island felt like a great honour.

When in a new location, I have few expectations, and little idea of what I may encounter. So the following morning, setting out with Sarah and Twomey in the dark, felt a foray into the unknown. The three of us tracked along the coastline with the continual roar of surf crashing on the outer reef, crossing small, sandy streams and negotiating around tangled stands of mangroves. Occasionally our torches picked out the eyeshine of crocodiles in the shallows.

Nearing the tip of the island, the path turned inland into the rainforest. Away from the pervasive roar of the reef, the forest was alive with scintillating insects and the soft barks of tree frogs. Here I suggested I'd try recording, and leaving Sarah and Twomey, walked a little further on to place my microphones. As I did so, I noticed a cricket giving an occasional trilling note, quite low in frequency. I mused that it must be a sizeable species.

With the recorder on, I returned to the others waiting quietly in the dark. With that conspiratorial smile that indicates she's about to share a secret, Sarah indicated I switch off my head torch. As my eyes adapted, I saw there were patches of luminous fungi on the forest floor. Growing on decaying vegetation, they were spaced irregularly, spread out in drifts here and there all around us. Glowing softly in faint points and clouds of light, they gave the impression of starfields. In the utter darkness of the rainforest, I imagined that we were standing among the nebulae of the milky way.

Entranced by this, I then noticed the cricket I'd heard earlier had been joined by another, calling at a slightly different pitch. Or was it a cricket? I asked Twomey, who confirmed my suspicion. "It's a bird," he said, "a flycatcher, little one, white head".

We were hearing White-capped Monarchs, an endemic species to these islands. I recalled seeing them around the hut the previous day, but not calling like this. Their tremulous whistles were now more frequent than earlier, and soon there were quite a number of them, roosting somewhere in the rainforest canopy. Each bird was giving a quavering note on a single pitch, repeating it every so often. Individually, their voices sounded ethereal, and in combination they were hauntingly beautiful. As each monarch sang its single note, floating it out into the night, it combined with others at slightly different pitches. Together, they formed tonal patterns, the notes of which I couldn't help hearing as being close to a musical scale. 🐦

The nocturnal singing of those monarchs remains one of the most beautiful and pleasing nature sounds I've ever documented. Here, in a distant corner of the western Pacific, we were being enchanted by another of those sonic patterns in nature that I so love hearing. Once again, like the Tawny-crowned Honeyeaters of the Little Desert, it was a community of one species, singing together with a shared repertoire, weaving hypnotic textures out of the simplest of sounds – in this case, single whistles.

I also suspect that we stumbled across a soundscape that has rarely been heard, let alone recorded. Even Twomey said he'd not been aware of the monarchs singing quite like this. Over following mornings we returned to the same spot to hear them sing. I understand that after we left, Twomey would encourage his guests to rise early and set out with him in the dark, to stand on the Milky Way and hear the heavenly monarchs of Tetepare's rainforest.

The Whistlers of Kolombangara

Sitting with Twomey on the verandah of his hut on the first day of our arrival on Tetepare, the three of us discussed what we hoped to achieve on the island. For context, I played him some recordings, possibly the first Australian birdsong he'd heard. As kookaburras chortled in the headphones, he tilted his head quizzically, "that's a bird?"

In the course of discussions, Sarah asked him what his favourite songbird was. Without hesitation he replied, "The Golden Whistler". Golden Whistlers are a familiar species in Australia, as they're quite common in eastern and southern forests, and frequent our woodlands at home. I recalled they occurred in the Solomons too, and consulted my bird guide to check. I showed the page to Twomey – *Pachycephala pectoralis*, the Golden Whistler – and commented that our birds looked similar but had a white throat, while in the Solomon Island's race it was a vibrant yellow. As their name suggests, Golden Whistlers have a beautiful voice, very sweet and rich, which I knew well. We could easily understand why it would be Twomey's favourite.

However during our ensuing days on the island, we didn't encounter a single whistler. This was not a surprise; Twomey told us they were usually found higher on the ridgelines, whereas we ended up concentrating our efforts on recording in the rich lowland rainforests. To be honest, I was completely entranced by the exotic encounters we were having with hornbills, coucals, Singing Parrots, Metallic Starlings, a variety of fruit-doves and those monarch flycatchers. On our final morning we watched as what must have been several hundred frigatebirds drifted serenely overhead, lit by the sunrise. By the time we waved goodbye to Twomey and pushed out from shore, I'd forgotten all about the whistlers.

From Tetepare we travelled on to Kolombangara Island, another two days crossing the waters, this time calmer and with an engine that kept puttering along. As we rounded a last shoulder of land, Kolombangara appeared ahead, a perfectly symmetrical volcanic cone emerging from the tropical sea.

Our base for field work was a sturdy forest lodge, perched at 400 metres altitude on the flank of the volcano. Its large deck overlooked the primary rainforest which stretched upward to the imposing caldera on the skyline. From the side of the lodge, an obscure foot track headed off into the forest, the path that led toward the peak.

Once again, we met the man appointed to be our local guide. Moffat was sparkier than Twomey with a cheeky demeanour, but as we went through a similar orientation, we felt he was equally knowledgeable. Sarah once again asked his favourite songbird. Moffat's reply; the Golden Whistler. I said we'd not heard one on Tetepare. "You will hear one here," he said emphatically. "Really?" I wasn't doubting, just curious. "Yes, tomorrow you will hear them."

The following day we arose at 3 a.m., and donned boots for our first taste of Kolombangara's rainforest. Setting off, we found the track slow going as the ground was uneven, volcanic rock, and the way had to be negotiated around trees, mazes of vegetation and fallen debris. The jungle was so dense and tangled that moving off the track was quite impossible. All about us, the night was alive with nocturnal sounds; a susurration of crickets punctuated by the sharp peeps of tiny frogs which seemed to be everywhere on the forest floor. 🐦

Following Moffat, we hiked uphill for nearly an hour, by which time, despite the cool of the night, I was unpleasantly sweaty. I called a halt, anticipating the approach of dawn and the transition from the nocturnal choir to birdsong. The jungle was quite uniform, so where we were seemed as likely to be a worthwhile spot as any other. Mounting the microphones and plugging in cables, I was startled by a massive stick insect landing on my face, presumably attracted by the light of my headtorch. It clambered scratchily around my forehead as I finished setting up. I chuckled; Sticky the insect meets Mr Sticky the sweatball. Ah, the discomforts and wonders of tropical forests. Delicately, I relocated the gangly, fumbling insect to a nearby shrub, and stepped back to join Sarah and Moffat standing quietly in the darkness to await the dawn chorus.

But the dawn chorus didn't happen.

As the first pale light emerged in the east, the calm was swept aside by an extraordinarily loud whistle; a banshee siren that quickly swept up in pitch, before nose-diving in stutters. Again it came, wild and exuberant. It was a bird, but I'd never heard anything like it. 🐦 Turning to Moffat, I asked what it was. "Golden Whistler," he whispered back. Now I really was doubting, wondering briefly whether he knew his birds after all. Moffat grinned back, enjoying my incredulity. Actually, I was stunned. This was utterly unlike the beautiful, sweet whistler songs I was familiar with in Australia.

But the banshee siren was only a prelude. Now the bird was settling into a series of percussive 'chop!'s, which cracked through the trees, filling the air with sound. It was so loud, my eardrums were distorting with each whack. They came in patterns; a few quick cracks together in succession, interspersed with piercing whistles and whipping inflections. Every now and then, further banshee whistles were interjected.

I squinted at my recorder a few tens of metres distant. Damn! I could see my usual record levels, set from years of experience so as to rarely overload, were now flashing into the red. Moving to wind them back, I was amazed at how much I had to attenuate before the bird no longer overwhelmed the inputs.

If this whistler had given only a few loud calls I'd have been wowed by its vocal ability, but it didn't stop. Moving periodically to another treetop, it kept singing, a torrent of high intensity sound that gave no indication of ending. As daylight arrived, we continued to be assailed with its astounding performance. Meanwhile, the only other bird we could hear was another whistler giving a similarly extrovert vocal showcase some distance away. Nothing else was getting a look in. Our whistler's assertive and staccato rhythms were infectious, and Sarah and I broke into a spontaneous dance, improvising bad disco moves in response. Moffat watched on with what seemed a mixture of amusement and concern for our sanity.

After nearly two hours of continuous vocalising, the whistler subsided, its song becoming increasingly occasional. The forest once again quietened, and in a mood of elated wonderment, I switched off the recorder. This Golden Whistler was now top of our birdsong playlist too.

That morning, and over subsequent days recording the whistlers on Kolombangara, I was not only awed by them, but intrigued. I had a gnawing inkling that I was hearing more than just a spectacular vocalist. The differences between the song of these Solomon's Golden Whistlers and the birds I was familiar with from Australia were close to total – I could find little similarity. Listening to them in this exotic jungle on a remote island gave me a strange feeling, which I couldn't place at first. As the days went by, I had an eerie sense of the unfamiliar, of being on the edge of something vast and mysterious. I had a growing realisation that these birds were an audible expression of a process we think we understand, yet have no sensory experience of.

I wasn't just hearing birdsong, I was hearing the process of evolution.

Chapter 6

Listening to Deep Time

We're not naturally gifted to appreciate Earth time. Our lifespans, a mere blink of time's eye, constrain us, and while we can intellectually understand the vast periods of evolutionary time, our ability to actually experience it is understandably limited.

But we can hear it – if we know what to listen for.

The process of speciation, the emergence of new species, is often thought of in terms of physical traits. In closely related species, these may be as slight as variations in body markings, plumage patterns or body size. In insects, the specifics of genitalia are frequently cited as being diagnostic. Focusing on these observable characteristics has made it easy for the writers and illustrators of natural history books, but as we've noted, this has slanted our conception of the evolutionary process toward the visual.

Sound is an equally, if not more, significant indicator of speciation.

Any two closely related species, and ultimately lineages of species, were once one. Hence at some point in the past, they shared a repertoire. As populations of that ancestral species began a process of divergence, through whatever circumstances of geographic isolation or adaptation to new opportunities, they gradually acquired different characteristics.

A unique repertoire is an important component of that suite of characteristics. For any soniferous creature, having vocalisations exclusive to their species will be essential. If members of a species are going to attract mates, bond, synchronise communications and negotiate interactions through sonic signals, they require a signature repertoire to do so. They need to 'speak the same language'. It is necessary to recognising their own kind and reducing any likelihood of re-mingling and hybridisation. So the emergence of a unique repertoire is no arbitrary process. It is central to the essence of a species, no matter what other adaptations may eventuate.

As vocalisations are relatively malleable, either genetically or through learning, we could expect that changes in vocalisations may be one of the signs pointing to the onset of speciation. Indeed, repertoire change may precede other physiological or behavioural adaptations. When divergence has happened in the relatively recent past, we may even encounter two species that still appear and behave quite similarly, yet sound significantly different.

Sound and Speciation – The Wedgebills

As one would guess from their names, the Chirruping and Chiming Wedgebills are most easily distinguished by their songs. These two species are found throughout the inland of Australia in two separate populations, one species in the east and the other from the centre across to the far west coast. They look the same, being of uniform brown plumage and sporting a rather cute, floppy crest atop their crowns. They frequent similar habitat, often being encountered in acacia woodlands and showing a preference for claypans and dense lignum scrubs. They both feed either on the ground or among low shrubs, and have the same shy and wary character, quickly retreating to cover if disturbed. Their breeding biology is identical, even to their behaviour of singing atop a low bush, among the few Australian birds to habitually do so.

It is their songs that are unmistakably different.

The song of the Chirruping Wedgebill was among the first I recorded, as they were reasonably common in the open scrublands around Mutawintji. It is indeed 'chirrupy', each identical phrase being composed of two, soft, lead-in notes preceding a louder, sparrow-like 'chirrup' syllable, and all sung in a looping cycle; *tutsi-CHEER, tutsi-CHEER, tutsi-CHEER* … Their repetitive songs are delivered in bouts lasting a minute or so, with an interesting feature of beginning softly and steadily increasing in volume until abruptly stopping. This is the male singing, but he is occasionally joined by the female who responds in duet with a quick, upslurred *whEET*. 🐦

Recording them on the open plains as they perched atop low shrubs, I thought their songs pleasant, unusual, but a bit predictable. So I wasn't particularly mindful of their companion species, the Chiming Wedgebill, until I encountered them a few years later.

Leaving Adelaide for our first field trip to the deserts of central Australia, we drove north. By the evening, after a day of travelling across a vast and increasingly arid landscape, we were tired. Noticing a dirt road heading off into the scrub, we left the highway in search of a quiet place to camp for the night. In the headlights, the country looked dusty, flat and monotone. I had in mind to put some distance between us and any noise from the highway, even though the country didn't look inspiring for recording. Some twenty kilometres down the track, we pulled over and rolled out our swag under a starry sky. It may not have been a particularly exceptional spot, but it would do for the night.

The following morning I rose in the early hours to an unexpected chorus of sleepy birdsong. The moon was up, and the calls of Magpies, plus Spiny-cheeked and Singing Honeyeaters drifted on the cool night air. As first light arrived, I found myself among a cluster of acacia bushes – perfect habitat, I soon realised, for Chiming Wedgebills.

As I heard the first one, I was entranced. Chiming Wedgebills also have a cyclical song, but unlike the burred notes of 'chirrupers', the voices of 'chimers' are like tiny bells. Cascading over and over in a descending series of silvery notes, the effect of their song is completely delightful. They have an ethereal vocal quality, which the oft-quoted onomatopoeia of 'did you get drunk' doesn't really capture. 🐦

Soon more Chiming Wedgebills began singing, and I realised I had stumbled upon quite a community of them. The effect of their voices interacting was to create carillons of delicate tones.

At times, the hypnotic cycling of sounds from left and right created a noticeable phasing effect, which for me listening among them, was almost psychedelic.

It was not only the Wedgebills serenading that morning. The whole landscape became alive with birdsong. Honeyeaters, Variegated Fairy-wrens, Mistletoebirds, Crimson Chats, Crested Pigeons, Western Bowerbirds, Cuckoo-shrikes and Mulga Parrots called, while Budgerigars chattered in abundance. Later, as I was returning to camp, a pair of Emus strolled out of the scrub and ambled alongside me for a way. This place was vibrant – so much for it not being exceptional.

Then I noticed a swathe of damp ground, a muddy washout. Further on was the remains of a silty pool now drying in the desert sun. Evidently it had rained here – and recently, I estimated within the last week. The surrounding country was bone dry, so whatever storm had dumped the rain must have been both heavy and very localised. Now the abundance of this otherwise nondescript location was explained. So many of Australia's inland birds are opportunistic breeders, responding to rainfall. Guided by some mysterious sense, the birds had come, possibly from some distance away, and I had stumbled into a very active breeding event.

The songs of the two Wedgebill species show their close evolutionary relationship. They both have a similarly cyclical pattern, with song bouts lasting up to a minute. They share that curious feature of beginning quietly and building to full volume before abruptly stopping, pausing awhile before beginning again. They also both prefer bushes as ideal song perches, and I watched as individuals moved from the top of one to another for their performances.

While the vocal habits of the two species show affinities, it is the actual songs that are distinctively different – one having that rough chirrup and the other clear, bell-like notes. Like oil and water, their voices have separated and identify each as unique.

But why have these species differentiated at all? Many years later, we travelled the north-south line where the two populations of Wedgebills could possibly meet and even overlap. I was curious about whether there would be any hybrid vocalisations, and what they may sound like. For a week, we crisscrossed the region of overlap, without hearing a single Wedgebill. The reason was obvious – it was one of the driest regions we've travelled through, made worse that year by a protracted drought. One area was so desolate and devoid of vegetation as to have earned the name the Moon Plain.

Looking at the bare expanse of country, it was apparent why there are two Wedgebill species. The drying of the continent, particularly at the end of the last ice age, has created a barren divide between east and west. The original Wedgebill population, which may once have inhabited much of the inland, has become separated by a zone of harsh and unsuitable habitat. In this geographical separation, they became first two populations, and then two species. It is estimated this evolutionary fork in the road occurred in the last few tens of thousands of years, and possibly only since the end of that last glacial age – not long by evolutionary timescales. Of all the physical characteristics that evolution could have acted on, voice has been the most pliant. Two of the most colourful sounds of the Australian outback, each being characteristic of the east and west of the continent, are the result.

Songs of Divergence –
The *Phylloscopus* Warblers

Wedgebills are far from the only species that show vocalisations to be a definitive trait that can evolve rapidly.

At the beginning of our European journey, in the garden of Istanbul's Blue Mosque, I recorded a Chiffchaff giving the song after which it is named. Later in that trip, I found the spring forests of Sweden filled with the sweet songs of Willow Warblers.

These two species, both warblers of the *Phylloscopus* genus, are closely related. Their plumage characteristics are so similar as to make them difficult to tell apart visually. This could be a challenge for observers, as they share similarly wooded habitats and co-exist in overlapping ranges that take in vast areas of Eurasia. Yet, despite these overlaps of geography and appearance, their voices clearly tell them apart. Even as a novice to Europe, I could never confuse these two small, green birds when listening to them. Identical twins to the eye, but to the ear, a whole different beast.

The story doesn't end with them. The Iberian Chiffchaff was formerly thought to be a race of the Common Chiffchaff, but has now been given full species status. This was long suspected based on – you guessed it – unique vocalisations, and now confirmed by genetics. Further afield there are the Siberian Chiffchaffs, Canary Island Chiffchaffs, and Caucasian Chiffchaffs. Some minor plumage variations become noticeable among these related species, but unique vocalisations remain a confident means of identification.

These *Phylloscopus* warblers are themselves embedded within the Palearctic warbler supergroup. Covering the expanse of Europe, Asia and the Orient, north of the Sahara and Himalayas, they number upwards of 80 species, many of them being similarly small, green (or brown) birds. These warblers can pose a challenge to visually identify in the field. The two weeks we spent recording in the Himalayan foothills were hopelessly insufficient for me to become acquainted with the dozen or more *Phylloscopus* warblers that occurred there.

Research is continuing into this large group of ambiguous birds, many of whom share close evolutionary affinities. The geological uplift of the Himalayan Range over millions of years has created plentiful opportunities for populations to become isolated, adapt to different habitats or ranges of altitude, and become new species. For researchers, vocalisations have become a key parameter in picking out possible speciation.[26]

And it's not just birds. Recently a population of Blue Whales off the coast of Sri Lanka was recognised as having a notably different repertoire. Subsequent studies have shown them to be a discreet population with their own feeding and migratory patterns, and possibly a new species.[27]

Meanwhile fruit flies, probably the most intensively observed group of organisms in lab studies, speciate in human time. Their rapid breeding rate allows researchers to witness change over multiple generations. Once again, it is their vibrational signals that show early and definitive variance.[28]

So from the small to the largest organisms on the planet, creatures are telling us who they are, who their relatives are, and to an extent, how far they have journeyed along their own evolutionary paths. In their everyday vocalisations, we are eavesdropping on the history of life.

Lineages of Sound – The *Petroica* Robins

If we can hear differences in song between closely related species, can we follow the progress of acoustic divergence through an entire lineage of species?

Australia has five species of robins in the genus *Petroica*, each having either pink or red-breast plumage. They are beautiful little birds, bringing an unexpected flash of colour to Australia's woodlands and forests – the Scarlet Robins that live in our bushland are one of them. These *Petroica* robins are a subset of a wider Australasian robin family (no relation to European robins), which comprises other groups such as the yellow-breasted robins.

Phylogenic studies – the study of evolutionary history and the relationships between species – show the *Petroica* robins as having evolved as a lineage, and it is reasonable to think this has been in response to climatic and habitat changes.[29] Over geological time, Australia has dried out as the continental plate has drifted northwards. The great and ancient Gondwanan forests that once covered much of the continent, have retreated to highland and temperate regions of the east and south, as much of the country has gradually become drier. This is a perfect scenario for the emergence of new species, and the *Petroica* occur to me as a likely example of a lineage that have made sequential adaptations.

The males of two of the five species – the Pink Robin and Rose Robin – have rich, pink breast plumage. Both are restricted to dense, moist habitats that are remnant of Australia's ancient forest ecosystems. This suggests them as closest to the original type, and phylogenic analysis indeed indicates the more southerly occurring Pink Robin as ancestral. This suggests the Rose Robin as having differentiated later by adapting to the warmer rainforest types along the east coast.

Meanwhile, the ancestral 'red-breasted' robin may have moved into the woodland habitats created by warmer and drier conditions, in time differentiating into the three species we find today. Flame Robins breed in cooler, upland forests, migrating down to the plains in the winter season. Our year-round resident Scarlet Robins show a preference for drier woodlands, while the Red-capped Robin is Australia's inland *Petroica*, inhabiting a range of semi-arid, scrubby country across the continent. A transition between species can be seen in the vibrancy of their breast plumage, which has deepened from the original pink, first to the vivid orange of the flames, then the rich red of our scarlets, with the red-caps finally adding a spectacular red forehead flash.

So by genetics, habitat preference and appearance, we have a genus lineage from the Pink Robin (putting aside the branch Rose Robin species), to the Flame, Scarlet and Red-capped – from wetter and cooler, to drier and warmer. From ancestral to more recently evolved. What can be heard in their songs?

Each of Australia's *Petroica* robins share a common pattern of uttering a single song phrase at somewhat regular intervals. We've already heard this from our Scarlet; its *pi-pi-pri-tidi, prrri-tidi* song.

But let's begin with Pink Robins, which are found on both the southern mainland and Tasmania. These two populations have been isolated since the submergence of the Bass Strait landbridge at the end of the last glacial period some 8000 years ago, and this is noticeable in their songs. While slightly different, they're nevertheless structurally similar; a rapid *wit, tinke-tinke-tinke-tinke-*

tink. The bird utters it much quicker than you can say it, coming across as a short and pleasant tinkle. 🐦

Analysis reveals an introductory note or two, followed by each of those 'tinke' components being made up of two elements of a higher and lower pitch. Trilled off quickly, they create an alternating 'up-down-up-down-up-down' pattern of notes on two, clearly delineated tiers of pitch.[30]

On first hearing, the song of the Flame Robin seems slower and more deliberate. It has a similarly tinkly quality, but it's not as rapid – you can easily discern individual notes. Like the Pink Robin, the notes of the Flame Robin reciprocate up and down. This zig-zagging of pitch, combined with an almost triplet pattern, gives its song a lilting cadence. 🐦 Comparing the two, there are obvious similarities; the tinkling quality of voice, and that it is made up of those two tiers of pitch. But in the Flame Robin, the higher and lower tiers are less clearly defined and more entwined, as though being blended together.

On to our Scarlet Robin. Its voice is also tinkly, but those higher and lower tiers of pitch have now blended to the point of vanishing. The entire song is at an almost uniform pitch. Scarlet Robins have lost the ancestral 'lilt'. 🐦

These three *Petroica* robins all share a similar song structure; a soft introductory note or two followed by a quick tinkly trill. But as we move through the species, the tinkly trill changes from one composed of discreetly tiered pitches, to less so, to almost none at all. So far, so good; there appears to be a pattern of gradual sonic morphing. Now we come to the Red-capped Robin.

At this point, I thought my enquiry had reached an impasse. The Red-capped Robin's song sounds so utterly different to the other *Petroica* robins. Unlike their tinkly, elfin-voiced songs, the red-caps give what sounds to be a pair of hard, whirring notes. It's a curious sound, like a miniature telephone ringing, with a buzzy, spiky tone – nothing tinkly about it at all. 🐦

Examining the structure, there are two lead-in 'chit's, followed by a very rapid trill which makes up that whirring note, finishing with a final 'dit' at the end. Two of these phrases in rapid succession,

and we have the Red-capped Robin's song: *ch,ch,PREEE,dit,
ch,ch,PREEE,dit*. As with the other *Petroica* robins, it is given as a
repeated phrase over and over, but there the similarity ends.

Or does it? I went back to have a closer examination of the
Scarlet Robin's song utilising spectrograms. Remember we've
described it as *pi-pi-pri-tidi, prrri-tidi*, and noticed that the 'prrri'
element is a quick trill. Also that embedded discreetly within the
first part of the phrase is another tiny, proto-trill; 'pri'.

Listening to the scarlet's song it doesn't sound much like the
red-cap's, but closer examination reveals the similarities. The
tinkly *pi-pi-pri-tidi, prrri-tidi* has become the buzzy *ch,ch,PREEE,dit,
ch,ch,PREEE,dit* by accentuating the trilled components to form the
red-cap's dominant elements. Hidden within the song of the scarlet
are two inconspicuous trills which, over the course of evolutionary
divergence, have become the more pronounced elements of the
Red-capped Robin's song. Considering that these trills allow robins
to range distance to their neighbours, I wonder whether their
prominence in the red-cap's song may be because they need to
project further, the birds being dispersed across larger homeranges
in a more arid landscape.

Although the song of the Red-capped Robin sounds superficially
different to its relatives, the patterns of inheritance are there on
closer listening. With this in mind, when I now hear the subtle trills
within the songs of our Scarlet Robins, I am aware of the potential
for songs to morph with the emergence of new species.

I also reflect that as these two robins diverged, the cognition of
each would have evolved in parallel with their calls. At every point
along the way, their minds must have remained in lock step with
their communications. Perhaps we can relate to this – as human
language evolves, we are always finely attuned to contemporary
usage. In a similar way, I can imagine that, as populations of
birds gradually diverged, they took the nuances of their calls
and songs with them, their brains remaining in synch with the
gradual transformation of their own particular repertoires over
evolutionary time.

When we learn to recognise species by their vocalisations, our minds are parsing those same evolutionary pathways. We listen for the acoustic subtleties of what makes their voices unique, recognisable or different. In doing so, we're tracing the lines of sonic divergence, running our mind's ear along the ancient trails of evolutionary history.

Voices Adapted to Habitat – Rufous Whistlers

It was spring by the time we returned from Kolombangara. The bush around our home was coming alive with birdsong again. After being so immersed in an exotic soundworld, it felt strange, as it often does, coming home and hearing the familiar. I felt sadness at leaving such novel and intriguing experiences mixed with relief to be home.

Walking along our bush track, I was enjoying the welcome of accustomed voices, while still pondering what I'd heard in Kolombangara. I had an ear out for our local Golden Whistlers, possibly to reassure myself that they really did sound so utterly different to their tropical relatives I'd just encountered.

But they weren't to be heard. I surmised that, as Golden Whistlers are most often a winter visitor for us, they had already dispersed southwards as they do each summer. Instead I came upon a Rufous Whistler in full voice. While in the same family, Rufous Whistlers are distinct from the goldens, with a ruddy salmon brown underbelly rather than the glowing yellow. Its throat puffed out in exertion, this bird was perched on a branch below the canopy of an old eucalypt, and was going for it. A series of acrobatic song phrases poured out, one after the other into the sunny morning. Its bold, liquid whistles were occasionally interspersed with an explosive *ee-CHONG!*, or long sequences of *joey, joey, joey* ... that kept up for half a minute. They really are the most wonderful songsters.

Rufous Whistlers occur widely across continental Australia. I've recorded them in the outback, in subtropical Queensland, and around the wetlands of Kakadu in the far north, where they are common among the paperbarks that line the billabongs. Remembering them singing there, I reflected that the tropical habitats of Kakadu felt almost as exotic to me as Kolombangara. Thinking of the two, I realised both Kakadu and Kolombangara are actually a similar distance away from the temperate woodlands where I live – some 3000 kilometres. And yet, the Rufous Whistlers in that far northern extent of their range sing similarly to the one serenading me now in the southern spring sunshine.

Why would the song of one whistler species remain much the same over such a distance, while another be unrecognisably different?

When listening to the whistlers of Kolombangara, I had the distinct impression that, despite what my field guide was suggesting, I was not hearing *Pachycephala pectoralis* at all – they were telling me at full volume they were something different. Of course I'm far from the first to notice this. Reading further on my return, I found that taxonomists are puzzled by the relationships of *P. pectoralis* throughout the islands of the western Pacific. For some they are the one species with a complex of subspecies and races. For others, full species distinctions have been proposed, with the Kolombangara whistlers being nominated as *Pachycephala orioloides*; the Oriole Whistler. A lovely name, although I suspect it references the bird's golden yellow plumage rather than anything to do with its song, which is not remotely oriole-like.

Putting aside the matter of whether the island and mainland birds are indeed separate species (and my belief that they are), I nevertheless continued puzzling over their extreme differences in song. Why are the Oriole Whistlers so ear-bendingly loud?

The acoustics of the Kolombangara jungles seem the tangible explanation. Thickly foliaged, they absorb and diffract sound, soaking it up like a sponge. For any bird to be heard over the distances required to communicate, they need to ramp up the volume. It's their only way of penetrating the thicket. The whistlers were not the only ones doing this on Kolombangara – those tiny frogs were piercingly loud too. Even with their explosive calls, I noticed the whistler's voices were barely conveying a few hundred metres – not far, but evidently far enough.

There is a pattern here. Loud vocalisations are characteristic of heavily foliaged environments. In Australian rainforests, many bird species have evolved songs with similar 'chopping' and whipcrack elements to the whistlers on Kolombangara. Chowchillas, Scrub-birds, Logrunners, Whipbirds, Eastern Yellow Robins and Lyrebirds, among others, all have a penetrating quality to their calls. Notably, many of these are ground-dwelling species, that often call from low down among tangles of ferny vegetation.

This suggests that the sounds birds utilise are tailored to the acoustics of their habitat, in particular the vegetation structure. Seems likely doesn't it? Natural selection should at least ensure that a bird's vocalisations are appropriate to where they live. This supposition is known as the Acoustic Adaptation Hypothesis, and has been widely tested.[31] The results have been broadly supportive, but in some instances ambiguous, and Australia's Golden Whistlers may be a case in point.

Like the Oriole Whistlers, they are also forest-dwelling birds. Why then do they sing so much more quietly? The reason seems associated with them being only infrequently found in dense rainforests. They prefer to inhabit a range of woodlands and taller eucalypt forests, where they hang out in the mid-storey. At this height, between forest floor and canopy, among upright tree trunks, there is generally less foliage to absorb sound. In this acoustically transparent strata, Australian Golden Whistlers can adopt a more relaxed voice.

Rufous Whistlers provide an interesting counterpoint. In the tropical woodlands of Kakadu they sing in a manner recognisably similar to our birds at home. Those northern woodlands, with their paperbarks and ghost gums, are visually very different to our box eucalypt forests. Crucially however, they are not dissimilar in their acoustic properties. Sound is dissipated to a similar degree, and transmits over comparable distances. So the songs that enliven our woodlands work equally well in Kakadu. This is true across the continent, wherever Rufous Whistlers find their home. Whatever the particular tree species, sound broadcasts in those woodlands in a similar manner. No adaptation is required, and a characteristic repertoire unites Rufous Whistlers across vast regions.

This understanding of what shapes birdsong is one of context. It suggests that a bird's repertoire has evolved largely in response to its environment, specifically the physical acoustic properties of its preferred habitat.

Listening to our Rufous Whistler singing in the spring sunshine, I had another realisation. Our winter visiting Golden Whistlers sound more similar – in vocal character, tone and projection – to Rufous Whistlers than their Kolombangara relatives. Despite their evolutionary separation, Rufous and Golden Whistlers have converged on similar sounds appropriate to the woodland habitat they both share.

In the songs of Rufous, Golden and Oriole Whistlers, we can hear how each of these birds have arrived at voices suitably adapted to where they live. In both the differences and similarities between them, we can hear the acoustic influence of their native habitats.

Family Likenesses – The Papuan Whistlers

Papua New Guinea is not a place you can simply wander around the bush. All land is owned by local families, and so gaining permissions and support to be walking through traditional lands is a prerequisite. Any journey is thus an expedition, and I was fortunate to be offered the opportunity to join one. No independent trip, for me this was to be a rare occasion when I'd be working alongside colleagues – all friends and fellow members of the Australian Wildlife Sound Recording Group.

Led by Tony Baylis, we set off to experience the Huon Range in the island's northeast. Tony had been to the region over several previous seasons, and established a network of local contacts. The previous year, ecologist Sue Gould and her partner Rod Thorn (our tech support for the duration) had joined him, and hence knew what lay ahead. I was the uninitiated, as was David Stewart, one of Australia's most accomplished and senior nature recordists. Dave was then in his 70s, and we were all worried whether he would be up to the challenge – a concern he readily shared.

For trekking in New Guinea is not for the faint-hearted. As we flew into the mountains in a light aircraft, I had my first views of the kind of landscape we'd be traversing: dense rainforest, precipitous slopes, landslips and razorback ridges. Banking sharply into a final valley, we could make out our destination ahead, the village of Sapmanga, and a frighteningly short grass airstrip cleared against the shoulder of a hill. Gliding in over the village's banana gardens, the plane's stall warning indicator began chiming, and our wheels hit the ground with a thump. Engine roaring, we pulled up a few metres short of boulders marking the end of the strip. So far, so good.

It seemed the entire village had come out to greet the once-weekly flight, and watched on as we disembarked our gear from the plane's hold. We were introduced to the senior locals who Tony had worked with on previous trips. These were the landowners and village chiefs without whose assistance we could not have visited. After acquaintances had been renewed, and with guides, Luke, Jonah and Keshdi, plus porters assisting with our gear, we set off on foot for the cloudforests.

New Guinea can be described as a vertical landscape. On our first full day of trekking, we only walked a horizontal distance of around three kilometres, while ascending over 1000 metres, climbing ever upwards through primary rainforest. Following a slippery and muddy foot track that was barely discernible at times, through swirling mists and rain, we placed one step in front of the other for hours on end. Our native porters had forged ahead, skipping barefoot up the slopes with our equipment while we plodded behind laboriously.

At one point, I chose to carry on while the others took a break. Climbing alone, I reflected that I'd never imagined having the opportunity to visit New Guinea. It seemed such an exotic location

– so close to home while tantalisingly out of reach. And yet, here I was, moving slowly through this primal, mist-infused forest.

The flora and fauna of the Australian landmass and New Guinea share a close kinship. Many species found in New Guinea belong to families I knew well from home. There were local types of fairy-wrens here, plus fantails, honeyeaters, sittellas and kookaburras. The famed birds of paradise have riflebirds as Australian relatives. There were even representatives of the *Petroica* robins. And there were whistlers.

New Guinea has a diversity of whistler species, over a dozen types, many sporting golden yellow plumages. Considering this wealth of varieties, it is likely the region was the ancestral home of the *Pachycephala* genus, with species not only adapting to local environments, but radiating southwards to the Austral landmass and eastward into the islands of the Pacific. So I was probably at the geographical centre of whistler evolution, and the crossroads between *P. Orioloides* and *P. Pectoralis*. I wondered whether the local whistlers here would sound more like the Kolombangara birds, or those I knew at home?

Listening now, it was almost silent. I could hear no birds, only the drip of moisture falling from the canopy onto the forest floor. I also couldn't hear any voices from the others. I suddenly had a moment of panic. Was I still on the track? The path was so vaguely defined and I'd been off in my thoughts. Inspecting my surroundings, my anxiety was tinged with wry amusement that I could manage to get myself lost on the first day as soon as I was out of sight of the others.

Scanning the ground intently, I continued onward, following the merest hint of a path. Even if I was on a track, it may not be the correct one. Had I missed a junction? Then, relief. A crisply-defined set of toe prints in the sticky mud showed that a barefoot walker had passed this way, and considering the daily rainfall in these forests, it had to be in the last few hours. Reassured, I carried on. Shortly after, voices up ahead heralded our porters returning homewards, having dropped their loads at our base camp. They were doing the round trip in the time we were slowly making our way up. From there, it wasn't long before I arrived, followed by the others, at the rough forest clearing in which we were to stay the next few days.

We were deep in primary rainforest, our campsite chosen as a rare patch of relatively level ground near a hillside stream. On arriving, our guides promptly set to with their bush knives, cutting poles and lashing them together with vines to assemble a pair of building frames. These were covered with the only housing materials we'd brought with us; two large tarps. It was humbling to watch them achieve this with such easy competence, finishing just in time to shelter us from a late afternoon downpour. As the rain pummelled down, a carpet of ferns was laid for flooring, pole shelving for our supplies erected, a smoky fire established, and as a finishing touch, a small, forked branch whittled on which to hang our coffee cups. Home sweet home.

The following morning, I arose and set off independently in the dark, following the track for a distance farther on to find a secluded recording spot. The others would also be out recording, and I didn't want to get in their way. With the microphones running, I stood back to await my first experience of the birdsong of New Guinea's cloudforests.

Dew fell steadily from above, as it often does in the tropics before dawn. The air was alive with shrill, peeping trills which seemed to be a constant feature of the place. We'd been told that, whilst sounding like insects, they were coming from tiny frogs. Against the now paling sky, I could just make out small insectivorous bats fluttering overhead.

Then the first birdsong began, and with a loud whipcracking call, I was immediately transported back to Kolombangara. I was hearing a whistler, and the similarity of ringing, percussive song to the Solomon Island birds was unmissable. Like P. *orioloides*, this song was composed of a sequence of 'chop!'s, but in a fixed phrase with a repeating pattern. With less improvisational variety, it was using similar sounds, but in a consistent pattern. Each phrase began with several incisive outbursts, followed by a curious, soft fizzing sound, and concluded with three or four more explosive 'chop!'s. After the briefest of pauses, the whole phrase would be repeated identically, creating a cyclical song that – as with *P. orioloides* – went on and on.

I was hearing a Regent Whistler, *Pachycephala schlegelii*, a common inhabitant of New Guinea's mountain forests. Like the Kolombangara species, and as its name suggests, it is a regal species, its golden plumage so rich as to verge on tinges of orange. When singing, it puffs not only its throat but the chrome yellow feathers of its nape.

Meanwhile the dawn chorus was gathering and swelling around it, a mix of ground robins, ifrits, fruit-doves, fantails, jewel babblers and melampittas. Many of the birds here seemed to call in melodic or at least tonal patterns, and the blending of them all was spellbinding. Some time later, the chorus began subsiding, the Regent Whistler being the last bird left singing, its voice still echoing throughout the forest.

There was no mistaking the kinship of this Regent Whistler with the Oriole Whistlers of Kolombangara – the volume, loud 'chopping' elements, and intensity of display were common to both. That these characteristics were present in the Regent's song was to be anticipated. The cloudforest was as thickly vegetated as the jungle on Kolombangara's volcanic hillsides, and was absorbing sound to a similar degree.

Over the next few weeks of our expedition, I was to hear further whistler species, some of whom sounded more akin to Australia's *P. pectoralis*. As with the *Petroica* robins, I could discern associations that spoke of species lineages separated by geographic and evolutionary distance. However the New Guinean whistlers are a far more confusing group, their repertoires suggestive of a complex history shaped over millions of years by rising mountain ranges, sea level changes and resultant island isolations. Despite all the ambiguities, the voices of these whistlers shared a penetrating capacity, uniting them in a common response to the acoustic of their heavily vegetated environments.

Hearing a Nascent Species?

The actual process by which evolutionary change may give each
species it's song I had considered unknowable – until one sunny,
autumnal morning. Walking our bush track, I heard one of our
Scarlet Robins giving a song variation. In addition to its usual
phrase, it was appending a repetition of the second syllable group:
pi-pi-pri-tidi, prrri-tidi, prrri-tidi. I'd never noticed a Scarlet Robin
do this previously. I listened for hesitation or variance, but no, the
song seemed clear and purposeful.

The implications had my mind leaping. Here was a Scarlet
Robin giving a song with two very clearly articulated trills, rather
than the one-and-a-hint of its usual song. With this variation in
form, it was now much closer to matching the two-trill song
structure of the Red-capped Robin.

That our Scarlet Robin was doing this in autumn may be
significant. It's the only time of year I've heard this song variation,
and I've since documented it over several autumns. Why only in
autumn?

There is some suggestion among biologists that at the end of
the breeding season, the neural networks of the songbird brain
that facilitate vocalising become diffuse. This may be a prelude to
some species falling silent over non-breeding months. Their singing
centres atrophy over winter, with hormonal changes stimulating
a regrowth the following breeding season. This shouldn't be
understood as some kind of mental deterioration, a 'songbird
dementia'. Rather, a process of cognitive efficiency, a byproduct
of which are song variations and errant vocalisations which are
more likely to be heard late in the season.

And there may be another aspect to these autumnal song
variations. My sound recording colleague, the late Dr Gayle
Johnson, researched the singing behaviour of Grey Butcherbirds
for her PhD. Like Pied Butcherbirds, they have a specific repertoire
of song phrases and ornamentations each year. Gayle suggested
they 'stretch out' toward the end of the season, improvising new
phrases additional to their existing repertoire. After going relatively

quiet over the non-breeding season, they begin the next season with a new repertoire – one built on the improvisations explored the prior autumn.

I began wondering; could a similar process inform changes over evolutionary time? Perhaps late season song variations are instrumental in the process by which populations of birds, separated by circumstances of climate, geography or adaptation to new habitats, develop the novel vocalisations necessary for speciation. This fascinating possibility has me hearing our autumnal robins as another way in which evolution may be audible. After breeding is complete, I hear them straying into song variations that have the potential to become adopted among breakaway populations on their way to becoming new species.

As we've noted with the various *Petroica* robins, one can often recognise specific sonic features linking closely related species. Could late season song plasticity be a mechanism by which these kindred repertoires have diverged?

Song Maketh the Species – The Lyrebird

Toolangi forest in Australia's southeast is my favourite location to encounter Superb Lyrebirds. One of their preferred haunts is Myrtle Gully, a cool rainforested valley resplendent with tree ferns, bracken and a dense understory, from which emerge sassafras and myrtle, both ancient Gondwanan trees. Epiphytic mosses clothe trunks and branches in soft fuzz and spongy beards. Massive eucalypts reach upwards, their canopies seeming to merge with the sky overhead. The whole place feels ancient, exuding a sense of what Australia may have resembled in its distant, wetter past.

A walking path follows this valley upstream, and I set off along it with my recording gear. Littered with bark debris and sodden from small streams crossing it, the track degenerates to a quagmire in places, requiring detours into the bush. Frequently the fronds of giant tree ferns overhang so closely that I had to wend my way between and under them. On the edges of the track, I could see signs of the soil having been churned over, as though freshly raked. These scrapings showed lyrebirds had been foraging; balancing on one foot while the other clawed at the ground in search of invertebrate delicacies. But I didn't need these signs to know that lyrebirds were present – I could hear them.

The powerful voice of one came from up ahead, its song echoing between the trees. Another could be heard further off to one side. By the time I approached the first, he'd stopped and slunk away into the undergrowth. Seeing a lyrebird is a special event, but I was content just to listen, and hopefully record them.

The challenge would be in recording an entire lyrebird performance, which could last between ten and twenty minutes. To do that, I'd have to find one of a male's several display grounds; cleared areas on the forest floor from which he performs. Placing microphones near the edge of a display area, I could leave the recorder running until the male returned to his stage. I've heard spectacular recordings from colleagues done in this manner, capturing the bird at close range.

However I wanted to document the lyrebird's song in the context of its environment, with the trees and fernery echoing and softening its voice. So instead of needing to find its actual display ground, all I had to do was be in its vicinity – just close enough. This still took a bit of listening, walking back and forth up and down the track, and finding the singing locations of various males. I was in no rush. I knew it would be largely fortuitous to be in the right spot at the right time.

Which, in the end, I was.

About twenty metres off the path, I could hear a lyrebird quietly moving around, somewhere among the ferns. Shortly after, he began his display with a minute or two of curious and sharply percussive clicks, like two stones being repeatedly tapped together. Then, with a throat-clearing cluck, he was into the skilled mimicry for which his species is renowned, stringing together impersonations of one species after another. They were extraordinarily accurate and lifelike.

As his song flowed out, I was taking mental notes; a Pied Currawong, Crimson Rosella, Brown Thornbill, Yellow-tailed Black-Cockatoo, Pilotbird, and then a gallant attempt at a Laughing Kookaburra. Being birds that call in chorus, kookaburras pose a challenge for lyrebirds. They usually begin by imitating one voice, faltering at the point when other kookaburras would be expected to join in. However, particularly adept lyrebirds can imitate two kookaburras at once. This is not something I've personally experienced, but I've heard a colleague's recording that is both clear and convincing. This bird hadn't developed that skill, and gave up on kookaburras plural.

After a roll call of forest birds, the lyrebird interspersed one of its own species' songs; a powerful and melodious ripple of sound which resonated through the trees. It is among the loudest components of his repertoire, as though he's stamping his own signature on all that has gone before; copyright lyrebird!

Shifting slightly, I caught a glimpse of the displaying bird. While singing, he was also dancing. Bringing his white, filamentary tail feathers forward, forming a canopy over his head and body, he was quivering them urgently. It was this mantle of long, gossamer feathers that I could just see moving among the ferns.

The lyrebird's visual display is as important as his vocal performance. He has cleared his display ground among the dense fernery especially so that it can be clearly witnessed. From where I was standing he was largely hidden, but from above he would appear a shimmering, silvery apparition. It must be a mesmerising sight for any observing female looking down. His tail feathers trembling while pouring forth such a varied stream of song, he would be spectacular in both sight and sound.

The mimicry of the lyrebird has rightly brought the species fame. It is a remarkable capacity. Equally to be appreciated though, is the bird's stamina. In the peak of the season, male lyrebirds will perform from dawn until dusk, with only an hour or two to catch their breath and forage. This investment of both skill *and* energy speaks of how significant their display is to their biology. It also had me wondering.

As I listened to lyrebirds singing up and down the valley, the issue that puzzled me was not so much their vocal ability, as the time of year I needed to be at Toolangi to encounter them in full voice. It was the middle of winter. Lyrebirds, both male and female, will sing all year round, but it is during the cooler months that the males are in full display and mating takes place. This is an unusual breeding schedule for any bird. Daylight hours are low and temperatures cool, all limiting food resources. It doesn't seem a wise time to be starting a family.

And yet, hearing them in that rainforest gully on an overcast day, their voices carrying in the still air, it felt instinctively right. Pondering this, I realised the atmospheric conditions could not be more suitable.

Firstly, it was calm and cool, and as a result, sound carried effortlessly through the trees. This is important – lyrebirds have extraordinarily dynamic songs. They range from delicate clicks and soft buzzes to those outbursts of signature song, plus a few explosive screams thrown in for good measure. This extreme range of dynamics is best heard in an undisturbed acoustic, where the quiet aspects of his performance can also be heard. I could clearly pick up those stone tapping clicks and soft churrings from even moderately distant birds.

Secondly, the sky was uniform and sombre, as often in the cooler months. This even illumination would be perfect stage lighting for the male lyrebird's display. From above, his white tail feathers would stand out dramatically against the dull vegetation. They would attract attention as though spotlit in the gloom. The evenness of illumination would also be perfect to accentuate the shimmering movement of his filament feathers. The female would get a wicked show. Additionally, calm and overcast conditions usually last the whole day, giving ample time for neighbouring lyrebirds to take turns in giving extended displays.

Finally, apart from the lyrebirds, the forest was largely quiet. There were few other birds singing at that time of year.

I contrasted this with my knowledge of the forest in spring. In place of calm conditions, the warming season often brings more unstable weather. Storm systems come through – often a series of them – bringing winds, rain, and even late season flurries of snowfall. The clarity of the acoustic is appreciably compromised. When the skies do clear, higher sun angles create a harsh light, throwing deep shadows in exactly those ferny places on the forest floor where the lyrebird most needs to be observed. Cumulatively, these changeable conditions would decrease the optimal times for display. But most significantly, the forest would come alive with spring birdsong, as all the subjects of the lyrebird's mimicry return to full voice themselves.

I could now appreciate the lyrebird singing in winter as entirely appropriate, presenting the skill and dynamics of its song and dancing to best effect. Here again, it seems reasonable that the environment has shaped the bird's song. However in the lyrebird's case, I believe it is not only the acoustic of the habitat, but the conditions of season, atmosphere and light that have influenced his singing. This is because the lyrebird's song is no ordinary display. Shaped over vast time, most likely in response to female preference, the male has developed a performance preeminent in the avian world. It is so crucial to the species' existence that utilising seasonal conditions has become integral to its success.

It is possible that lyrebirds have always sung in the winter for some ancestral reason. However it seems more likely to me that they have come to do so in response to the optimal conditions of cooler months. If so, then we can imagine an incremental evolutionary process, in which the growing elaboration of the display has been paralleled by a drift toward the conditions in which it is both seen and heard most spectacularly. Suitably impressed and wooed, females have gradually been drawn to breed earlier and earlier, lengthening the period between mating and the time more suitable for hatching their single egg. All this has resulted in the lyrebird having the longest incubation period of any songbird, at over fifty days.

Listening to my lyrebird embody his surroundings in a stream of song, I felt the presence of vast time and slow change. I heard his performance as an interaction between his species and an environment that has remained relatively stable over evolutionary ages. Even Myrtle Gully itself feels like a place out of time.

Generation after generation, his ancestors have been refining their skills of mimicry, creating ever more virtuosic songs. In developing this prowess, they've also been drawn to seasonal conditions that have enhanced the success of their efforts – that calm time when few other birds are vocal. In the process, they've encountered a problem; that of breeding timing. One would consider this to be a physical problem that evolution would weigh as being more significant. Yet the display of the lyrebird has taken preference, and the birds have adapted their entire breeding biology in response.

My lyrebird abruptly fell silent, his performance concluded, and I imagined him shuffling his long tail feathers back into place before padding off invisibly among the ferns. Without his song, the forest seemed somehow bereft. I'm aware that for the lyrebird, of all the influences on survival that could take evolutionary preference, it has been sound.

The lyrebird's song has shaped the bird.

From this perspective, I could appreciate that in a way, the lyrebird has sung itself into being. Listening to the now silent forest, and knowing the lyrebird's ecological role in it, I reckon it is also singing Myrtle Gully's continued existence.

Chapter 7

Sonic Strategies

Vocal Learning

The Bundjalung Aboriginal People of Australia's central eastern coast and hinterlands have a story about the lyrebird. In the everywhen of The Dreaming,[32] the lyrebird was the first bird. When other birds appeared, they didn't have any songs of their own – it was the lyrebird alone who sang. When the lyrebird realised that all the other birds were songless, it gave each of them a song from its own extensive repertoire. When I first heard this story, I found it a delightful inversion of our understanding that the lyrebird appropriates the songs of other species to augment its own.

A contemporary perspective is expressed in the layout of many comprehensive bird guides. If you flick through one, you'll usually find species are arranged in a quasi-taxonomic order approximating the evolutionary development of orders and families. In the early sections you'll often find flightless birds, sea and water birds, fowl, raptors, then pigeons, parrots, night birds, cuckoos and kingfishers. Somewhere around the middle of the book, you'll turn a page and arrive at a single, extensive group which occupies the remainder of the guide. These are the *Passeriformes*, or passerines – which include the songbirds.

Passerines can generally be described as small, perching birds, often of woodland or forest habitats. They include many familiar types; robins, wrens, warblers, starlings, flycatchers, thrushes, finches, honeyeaters, crows and magpies, plus exotic groups such as birds of paradise, sunbirds and bowerbirds. This diversity shares a single physical characteristic in common. Whereas non-passerines sport a variety of foot types, such as the talons of raptors and owls, the webbed feet of water birds, or the two hind toes of parrots and cuckoos, passerines have three toes forward and a strong hind toe – ideal for perching.

Additional characteristics are specific to the songbirds, and for us as listeners, far more significant. The majority of songbirds have a complex group of muscles attached to the walls of the syrinx which gives them a very precise degree of control over their vocalisations. This allows them greater suppleness, expressive range and mastery of their singing than other birds can usually muster. To this physiology is added a cognitive development. While non-songbirds inherit their calls, and hence sing by instinct, songbirds must learn their songs. From being born with a general vocal character, they develop their specific songs by accurately copying parents and kin.

This capacity for vocal learning is relatively rare in the animal world. It is known among humans, elephants, bats, seals, whales and dolphins. The dissemination of new songs among a population of Humpback Whales is fine evidence of learning. While primates don't seem to have the ability, it is crucial to human development, as we know from infants.

In the avian world, vocal learning is evident in several groups; parrots, hummingbirds, the songbirds and possibly some ducks. It seems it may have emerged separately in each group, as they each employ it in their own unique ways.

That cockatoos and parrots should be vocal learners is appropriate to their intelligence and social lives. For captive birds, this sociability has them copying human voices around them, sometimes with uncanny accuracy and seemingly impish intent. For wild birds, their learned behaviours are often expressed in flock interactions. Watching as our local Sulphur-crested Cockatoos and Little Corellas wheel across the evening sky in loose mobs before descending into roosting trees for the night, is to see life lived collectively. Their loud screeching, followed by antics in the treetops, sometimes hanging upside down with wings extended, speaks of their playfulness, camaraderie and joie de vivre. While their air-peeling shrieks sound all of a kind to us, one nevertheless senses they're calling out their individuality; "Hey! Look at me!" 🐦

Budgerigars, those delightful small parrots of our inland country, may also gather in large flocks when feeding or drinking. As they collectively take to the air, they do so with an explosive rush of whirring wings and twittering calls. In the midst of this fast-moving cacophony, it is thought that individual birds are able to recognise their mates and offspring by voice alone. 🐦

While the vocal learning of cockatoos and parrots seems integral to their highly engaged social lives, it is not actually a prerequisite for individual recognition. Many non-learning water and sea birds are remarkably adept at locating their young or parents among noisy breeding colonies. In all these non-songbird groups, whether learners or not, I hear their expressive voices as conveying qualities such as character, intention and emotional state.

In contrast, songbirds may employ vocal learning for quite different purposes. For them, songs may be learned not so much to express individuality, as almost the opposite; to refine a very specific repertoire. Perhaps they are doing something similar to ourselves when we learn music – they have to practice to get it note perfect.

Young songbirds can occasionally be overheard polishing their act in a delightful process known as subsong. It is a privilege to hear them do this, as I get the feeling they're so totally absorbed as to be off in their own worlds. Often they practice from the security and anonymity of a dense shrub, and if a breeze is blowing to hide any flubs, all the better. In a quiet voice, as though whispering, they'll ramble through a great range of vocalisations. While mainly imitating what they've heard from their parents, they may occasionally include snippets from other species. They often seem to be improvising as well, as though exploring what their voices will do. I find it especially cute to overhear those louder aspects of their repertoire being practiced in a subdued voice that is barely audible. 🐦🐦🐦

At this stage in their young lives, songbirds are learning about the world and perfecting their communication. Like adolescent human minds, they are sponges for experimentation and refinement. They go through similar phases of development in which social stimuli are crucial to learning, and their audio processing abilities are integrated with their vocal skills. With subsong, they even have a 'baby babbling' phase. By the time they reach adulthood, accuracy of song will be crucial to integrating with their peers and bonding with a mate.

It's easy to underestimate the profound evolutionary step that vocal learning has been. It is as much a cognitive and neural development as a sonic one. For songbirds, vocal learning has opened a myriad of possibilities – new ways to be and survive in the world. When and where did this pivotal step take place?

Tim Low's fascinating book, *Where Song Began,* picks up the story.[33] For many years, it was believed that the archetypal songbirds of the northern hemisphere, such as thrushes, nightingales and wrens with their fabulous songs, had to have originated in their homelands. This seemed unarguable to northern naturalists, who viewed Australia's bird fauna as derivative, descended from relatively recent dispersions southwards via Asia.

As evidence of ancient origins for Australian songbirds began emerging from genetic studies in the 1980s, many researchers found it difficult to accept. The implications disturbed traditionally held theories. The picture that materialised was that the voices of songbirds had been echoing through the forests of ancient Gondwana well before any appeared in the north. It had been a radiation out of southern lands by which songbirds gradually colonised Eurasia and north America, not the other way round.

Thus, the world's contemporary songbirds are descendant from those first Gondwanan species. If you flip through a field guide to Australian birds, at roughly that middle point, you'll turn a page to the first of those songbirds, the contemporary families with the most ancient lineage. They are the lyrebirds and scrub-birds.

Yes, the Bundjalung Peoples are correct. The ancestor of the lyrebird was the first bird who gave the gift of song learning to all the other songbirds that evolved from it.

The First Songbirds

There are two lyrebird species. Superb Lyrebirds are the more well-known, and the ones that had accepted my presence at Toolangi. Albert's Lyrebirds are similarly skilful mimics, but have a more restricted range, being only found in mountain rainforests that are also – like Myrtle Gully – relics of Australia's moister past.

These ancient Gondwanan habitats are refugia for several other lineages of early songbirds whose descendants survive into the present day. In the hinterlands of southeast Queensland, on the highest crests of the range where age-old Antarctic Beech forests still survive, I recorded rare Rufous Scrub-birds. While searching for Albert's Lyrebirds in nearby rainforest, I was also encountering families of Logrunners. Eastern Bristlebirds live in these same upland areas, although in perilously low numbers. Meanwhile, much further south at Myrtle Gully, Pilotbirds literally live alongside lyrebirds, being named for their habit for following them and picking through their scratchings for leftovers.

The vocalisations of these species show two characteristics that are likely ancestral for early songbirds. The first is sheer volume. Heard up close, lyrebirds will frequently overload the eardrums. Rufous Scrub-birds are also extraordinarily loud – I've recorded their chipping song clearly above the sound of a cascading waterfall, and on another occasion risked my hearing trying to get a glimpse of one skulking in the undergrowth. Their kindred species, the appropriately named Noisy Scrub-bird, lives in similarly dense vegetation, although in coastal heathland where its voice contends effectively with the ever-present roar of surf. Logrunners and their northern rainforest relatives, Chowchillas, are also piercingly loud, as are the Pilotbirds, a favourite mimicry subject for Superb Lyrebirds. The sweet but penetrating song of Pilotbirds has a similar quality to that of the bristlebirds, several populations of which are also coastal heath dwellers.

These modern descendants of ancient lineages are all ground foragers, and it is reasonable to conclude their loud voices have been shaped by a life spent on the forest floor among dense vegetation. Worryingly, as those wetter habitats have contracted to relic areas and climate change dries them further, many of these descendants of the world's very first songbirds are now hanging on in small and highly vulnerable populations.

Meanwhile, other early Australian songbirds have adapted to drier habitats. Members of the treecreeper family are found around the continent, and we have two representatives in the bush around our home; the White-throated and Brown Treecreepers. Despite living in a more open acoustic, treecreepers haven't given up the habit of pumping out the volume. On occasion, I'll be listening to the pleasant twitters of thornbills, sittellas and robins only to be ear-slapped by the piping of a treecreeper nearby. 🐦

When I hear these contemporary species, I imagine the sounds of those very first songbirds. In their penetrating voices, I hear an ebullient echo of the avian inhabitants of that earlier, more densely forested world, and the evolutionary innovation that was beginning to fill it with song. That innovation – the other ancestral characteristic of songbirds – is the ability to learn through copying.

Mimicry

When song learning first evolved, the possibility seems to have propelled the emergence of new lineages of birds exploiting its potential. In addition to lyrebirds, the bowerbirds – another Gondwanan family descendant from an ancient lineage – have similarly woven skilful mimicry into a spectacular breeding display. During an energetic performance, the male bowerbird strings together his repertoire from jumbled, train-of-thought medleys of imitation, interspersed with his own species' songs. This manner of song construction is very similar to that of lyrebirds, and the kinship of behaviour seems too much of a co-incidence to have emerged entirely independently. Adding to this theme of ancient Australian songbirds, the scrub-birds also mimic, although more discreetly.

Before this apparently ancestral enthusiasm for integrating imitation into song displays, songbirds initially evolved a mechanism for exactly the opposite – for limiting mimicry. This was required so that species could learn and remain faithful to their own songs. By limiting learning to an adolescent phase, usually the first year or two of life, a young bird will perfect its song during a period of sensitivity when it is with its parents, in a process called close-ended learning.

Subsequently, evolution seems to have led to this limitation becoming 'relaxed', and many contemporary songbirds are thought to be life-long learners. Even lyrebirds and bowerbirds, which learn their mimicry repertoires largely from their kin rather than directly from other species, are known to acquire new sounds throughout life. Similarly other species adapt their songs in response to neighbours or the requirements of dispersal and integrating into new communities. This open-ended learning maintains song stability and is probably responsible for the remarkable consistency of repertoire that we hear among many songbird communities.

Mature age learning may also facilitate mimicry in many species. Of the four hundred or so Australian songbirds, nearly a quarter are known to mimic other species as part of their adult repertoire. Magpies are well-known and versatile at doing so. Many of Australia's warbler family, including various scrub-wrens, thornbills, the Speckled Warbler and Redthroat, are all adept mimics. Olive-backed Orioles frequently imitate woodswallow species, who are themselves skilful mimics. I've heard Chestnut-rumped Hylacolas mimicking Silvereyes, and Silvereyes mimicking Jacky Winters. One of the favourite subjects of Albert's Lyrebird is that other great mimic, the Satin Bowerbird. I sometimes wonder whether our birds ever get in a tangle over who's supposed to be copying who.

I find it exciting to recognise mimicry among a bird's repertoire. This is easy when a bird copies a familiar sound from the man-made environment, such as our Magpies are fond of doing. It is far more challenging to pick up imitation of other species, as only a fragment of song may be included. Being able to recognise these snippets, and who they're borrowed from, is one of the joys of learning to identify the birdsong of a local area. Listening out for mimicry certainly sharpens the ears. On rare occasions, one may even notice 'mimicry miles'; a grab of song from a species not found locally which has been learned far away, possibly by a migrating individual. Rarer still, from a species no longer found locally, the ghost of its voice being passed down the generations by a mimicking species.

Of course all this prompts the question of why birds mimic at all. For many species, it seems a means of acquiring what I've heard described as 'cheap repertoire'; an easy way of extending, ornamenting and adding novelty to a song. This may be to impress a mate, or for the cognitive stimulation of incorporating new sounds, or both. And there are additional uses of mimicry, one of which I'll come to in a moment.

As fascinating as mimicry is, it is only one of many possibilities that vocal learning has provided songbirds. From humble beginnings in the damp forests of ancient Gondwana, they have spread out to populate the globe. Exploiting a range of terrestrial habitats has certainly been part of their success story, but so too have been their innovative communications. Song learning has been a profound development, setting the stage for the evolution of a myriad new vocal behaviours, which have significantly contributed to the group's proliferation. The story of how it is done so is as complex as the various lineages and their evolutionary development, yet here again, a fundamentally new way of employing sound has changed the avian world, and with it, our planet.

Sonic Strategies – Songbirds

Each soniferous species utilises sound in its own way, bringing together physiology, repertoire, behaviour and communicative purpose into a coherent strategy for living. I think of this as a creature's life in sound – what we might call it's 'sonic strategy'. I find this a useful concept, as it helps us recognise the multitude of ways that nature has found to work with sound and achieve specific purposes. We can also appreciate why significant sonic strategies have informed the evolution of subsequent variations on a theme.

Songbirds, by learning their songs, have evolved a master strategy from which numerous secondary innovations have emerged. For an example of this, I'll return to lyrebirds and bowerbirds, which we've already noted share related forms of mimicry-adorned display. They also exhibit another very specific sonic strategy; responding to threats by mimicking birds of prey.

I recall the researcher Syd Curtis, who had a great affection for Albert's Lyrebirds and spent years recording them, telling the story of one of his early recording attempts. He'd placed a speaker near a display area to play back a recording of a rival bird. No sooner had he switched on the speaker than the resident lyrebird flew directly from the other side of the valley, and standing at the edge of his area, poured forth imitation after imitation, all being birds of prey species. He kept this up for twenty minutes, and Syd reflected that, although he got a cracker of a recording, the agitation induced in the bird had so distressed him that he'd never used playback again.

Bowerbirds also imitate raptors under very similar circumstances. On the bank of a dry creekbed in central Australia, I'd set up my microphones near the display bower of a Western Bowerbird. Over several hours he'd had no female visitors, and hence done little more than rasp and hiss occasionally. However when I returned to retrieve my equipment, he suddenly began imitating not one but two raptor species – a Brown Falcon and Whistling Kite.

I don't know how lyrebirds and bowerbirds have come to mimic birds of prey in response to threats. Is it simply a result of natural selection acting blindly, or are they aware on some level of what they're doing? Whatever the answer, the observation remains that these two related bird families share significant aspects of their vocal behaviour in common. They weave skilled imitation into similarly structured songs as a sophisticated mating display, and mimic raptors at their display grounds for defence. We can understand them as sharing closely related sonic strategies.[34]

Another specialised sonic strategy is found among Australia's miners. The genus *Manorina*, comprising the four species of miner, is part of the larger honeyeater family, the *Meliphagidae*. Many of the larger honeyeaters, especially the wattlebirds, have a habit of being pugnacious, however the miners have taken this tendency and ramped it up to ten – or eleven in the case of Bell Miners.

Miners share a behaviour of acoustically mobbing intruders. The Noisy Miner is well-known to urban dwellers in the east for its high-pitched chorusing of insistent *wee, wee, wee, wee* ... A colony will collectively harass and intimidate anything encroaching on their domain – other birds, predators and humans alike. This irritation factor succeeds, as I can testify having grown up with them in suburban Sydney. ♪ The Yellow-throated Miner uses a near identical behaviour across its continent-wide range. Tellingly, its very closely related cousin species, the Black-eared Miner, now reduced to a tiny remnant population and highly endangered, is relatively more demure.

But it is Bell Miners that have taken this sonic strategy and run with it. Also living in colonies, they create a sonic exclusion zone around their patch of forest. With their incessant, loud, chiming calls, from daybreak ♪ to nightfall, a Bell Miner colony will deter any other forest bird from entering their domain. It is a constant wall of sound, functioning to dissuade others. ♪

Bell Miners are being obnoxious for good reason. The lerp they feed on is the exudate of small insects. A rich, sugary substance, it is a nutritious food favoured by other birds. By keeping an area of forest to themselves, Bell Miners effectively farm and harvest the lerp exclusively, eating the sugary exudate but leaving the insect alone. A downside to this is that, in the absence of predators, psyllid insects can proliferate in the areas of forest colonised by the miners, sometimes killing off trees in a process known as Bell Miner Associated Dieback. There are suggestions that human forest fragmentation, logging and subsequent lantana infestation may be worsening a problem not so prevalent previously. Nevertheless, it is ironic that the sonic strategy that is successful for Bell Miners can come to have a detrimental impact on a forest's ecology as a whole.[35]

There are many other sonic strategies among songbirds, some of which we've encountered previously. The *Petroica* robins utilise phrases with embedded homeranging information. Magpies and butcherbirds duet with complex vocalisations to both affirm family relationships and communicate to neighbouring clans. Choughs use individually expressive vocalisations among their group, while employing common repertoire for alert and bonding calls. Fairy-wrens sing before daybreak to facilitate mating outside their familial kin.

I could go on, but I hope you get the idea, and can appreciate the significance. Sonic strategies – ways of utilising sound – enable unique possibilities for living. As with the lyrebird, a successful strategy may have a powerful influence, shaping other aspects of a creature's existence. From an evolutionary perspective, while we think of the creature making sound, the sound can also make the species.

The kind of vocal learning found in songbirds is universal to the group and a prerequisite for the range of communicative behaviours we find among them. It must have played a significant role in their success story, in why songbirds have proliferated and diversified.[36] But what about the non-songbirds, those that don't learn, instead inheriting their calls genetically? Is the principle of sonic strategies applicable to their lives too?

Sonic Strategies – Non-Songbirds

The place seemed deserted. Sarah and I stood beside our backpacks on the edge of a rural road, squinting into the mid-morning sunlight. We'd arrived. The tropical rainforests of Sulawesi's Lore Lindu National Park clothed the surrounding hills, and a massive fig tree overhung the road. As our taxi driver gave us a last, cheery wave and departed back to the now distant metropolis of Palu, we took stock. Our destination, the village of Sidaunta, was revealed as a handful of nondescript wooden and iron sheet buildings lining a bend in the road. It was quiet, with no one in sight. While having enough experience of Asian travel not to be concerned, we nevertheless glanced around with a slight feeling of abandonment.

A few days previously, we'd met with forest department officials in their Palu head office, and this village had been suggested as a good base to access the rainforest. On a map, they'd pointed out a walking path which departed the road at this point, winding up through the rainforest to distant villages. Without much more planning than that, we'd set off.

A teenage boy now emerged from one of the buildings, but before we could call out, he hopped on a motor scooter and rode off. The sound of his bike disappeared into the distance, and silence fell again. As we debated what we might do, a young woman emerged from a house opposite and walked over to us. After a brief exchange in our basic Indonesian, she indicated for us to follow.

Walking in from the bright sunlight, we were guided through the house to a large kitchen area at the back. Here we were introduced to an older woman obviously familiar with being in charge. This was Ibu Ira, the woman's mother and head of the household. Ira's English was even more fragmentary than our Indonesian, however we understood that we were welcome, indeed expected, to stay in her home. We got the impression from peals of laughter that we'd made her week; two westerners turning up so unexpectedly. We were straining to understand not only due to language limitations, but because there was a waterfall roaring in Ira's kitchen. Along the back wall, a continual flow of water, channelled by way of bamboo troughs, cascaded into a concrete basin. I guess that if you have a hillside stream flowing past your back door, it is a sensible way of plumbing your home.

We'd fallen on our feet. The small forest track that'd been pointed out on the map came out right beside Ira's house. Enterprisingly, she'd set up some benches for commuting villagers who'd walked down from the hills to rest as they awaited the morning bus into the Palu markets. From this captive audience, she made a modest living supplying refreshments and meals. And now she had some lodgers. A room was cleared, thin mats laid out for us on the floor, and we settled in with our new family.

Rising a few hours before dawn the following morning, I set off on my own to explore the track from Ira's door. Sarah planned to follow once there was light enough to photograph. In the dark, the tropical air felt close and saturated. By torchlight, droplets of humidity were suspended in drifts, and condensation fell continually from the foliage above. This patter of dewfall combined with a soft chorus of nocturnal crickets, and as I walked, I'd catch glimpses of glow-worms winking from earthen banks and nearby vegetation.

For the first kilometre or so, I passed through village gardens of bananas and coffee, but eventually my surroundings transitioned to the unbroken canopy of primary rainforest. I wanted to get as deep into the forest as I could before dawn, so kept going, following the path as it wound higher, hugging the side of a valley past small tributary streams and out around hill shoulders.

An hour or so later, it was a sudden, deep booming sound in the dark that halted me in my tracks. I froze, part puzzled and part anxious that it was something threatening, while also fumbling to get my recorder running. Silence. Then it came again; two notes, deep and resonant. 🐦 I had no idea of what I was hearing. Could it be a macaque, a monkey of these forests? Or a cuscus – would a cuscus make this kind of sound? I doubted it. As the minutes crept on, the resonant calling settled into a pattern, coming every fifteen seconds or so. Eventually, birdsong began to filter through the trees, and I could glimpse the first hints of approaching daylight. Suddenly, a slapping of wings above me signalled a pigeon taking off, flying swiftly away through the trees.

So – what I'd been hearing was a pigeon, one with a voice so deep as to have me wondering whether it may be a significantly-sized mammal. It turned out to be a White-bellied Imperial-pigeon, a species endemic to Sulawesi, and the largest pigeon occurring there. When I later spied one in the canopy, it was a handsome green and white bird, about the same size as the magpies we have at home.

This comparison points to the significance of pigeon calls. It may seem obvious that the lowest frequency a bird can produce is determined by its body size. The bigger the species, the larger its vocal anatomy, and hence the lower the frequency it can generate. While its vocalisations may extend into higher frequencies, this lower limit is set by physiology. Thus, bigger birds have more full-bodied voices while smaller birds tend to have higher, twittery ones. Comparing a diversity of species from large to small, and plotting their body size vs lowest vocal frequencies on a graph, results in an ascending baseline that unites them all.[37]

Except pigeons. For their body size, pigeons call nearly two octaves below what one would expect. While our magpies have a vocal range down to around 500Hz, this imperial-pigeon was calling at 150Hz. If one of our magpies ever produced a bass baritone sound like that, I'd be seriously raising an eyebrow.

How do pigeons create such low frequencies? This puzzled me for some time, and it is only relatively recently that I've found the answer. Most birds sing with mouths open, often with heads thrown back and throat quivering as they form and project sounds by manipulating mouth cavity and tongue. A pigeon does something visibly different. Calling with bill closed, it compresses its head down into its body and balloons its chest, as though pumping a bellows. It turns out that while creating vibration with its syrinx as normal, by shutting the mouth and nasal passages, a pigeon redirects each exhalation into its oesophagus. In birds (also amphibians and reptiles, but not mammals), the oesophagus is lined with smooth muscle and capable of significant expansion. Pigeons employ this property, each breath inflating the upper oesophagus into a resonating sac, its call then being heard radiating out from under the skin. Not only does this mechanism of vocalising lengthen the airway to produce an unusually low pitch (like a trombone), but it effectively filters any higher harmonics, resulting in that ventriloquial quality characteristic of pigeons.

This very different way of generating sound is known as a 'closed-mouth vocalisation'.[38] It is a means of calling that is found throughout the pigeon and dove family. Tropical imperial-pigeons are at the larger extent of pigeon size, while the dainty Peaceful Doves in our bushland 🐦 and diminutive Namaqua Doves of eastern Africa 🐦 are among the smallest. Yet each has calls that are produced in this manner, and are far lower in pitch than would be expected for their size. Indeed, adding pigeons to our graph of body sizes vs vocalisations results in another neat baseline, paralleling other birds but at a lower frequency.

Pigeons are non-passerines. They inherit their vocalisations. As we've surmised with our bronzewing pigeons, their hard-to-locate calls are an effective defence strategy. Hearing that imperial pigeon in the dark, I could not discern where its voice was emanating from. Closed-mouth vocalising, being universal to pigeons and doves, is likely ancestral. Hence it can be supposed to have benefitted pigeons ever since the earliest members of their group evolved, most likely in tropical rainforests. That pigeons have proliferated around the globe shows how successful they've been. All the while, their remarkable sonic strategy of deep calling has been retained. It is not an accidental trait that has followed them, but integral to their success as a modern bird family.

Pigeons and doves are not the only bird family to produce size-defying, low frequency calls. In our bushland at home, we have a species that is a third the size of our bronzewing pigeons, yet calls at an even *lower* frequency. Beginning almost inaudibly, a steady series of booming notes, *ooom, ooom, ooom ...* will grow in intensity and drift slightly higher in pitch over half a minute before abruptly stopping. It is always a delight for me to hear this call, because I know that one of the rarest and most cryptic species of our woodlands is around – Painted Button-quail. 🐦

While a bird that swims like a duck and quacks like a duck is probably a duck, a small, dumpy, brown, speckle-plumaged bird that looks like a quail and forages on the ground like a quail may not be a quail at all. The button-quails – sixteen species worldwide with half of those found in Australia – are their own unique family and unrelated to true quail.

Here again, sound is their distinguishing characteristic. While true quail have whistling calls, the button-quail have booming voices, produced in a form of close-mouth vocalising made possible by an enlarged trachea and an inflatable bulb in the oesophagus. Also, button-quail are an ancient lineage, and like the Emu, it is the female who is the vocal one, inflating her body dramatically as she calls. Hearing her in our bushland, I experience that same sense of the ventriloquial as I do with pigeons. It seems an impossibly low call for such a tiny bird, and I'm confident it plays a similar role: a way of communicating while remaining cryptic.

In addition to pigeons and button-quail, closed-mouth vocalisations are found across more than a dozen different bird groups, for instance being responsible for the booming calls of bitterns, bustards, cassowaries and – yes – Emus. As these families are mostly unrelated, it seems that each have separately evolved ways of creating low frequency sounds, suggesting just how significant the advantages of this sonic strategy are. While locational ambiguity is likely to be one benefit, the calls of Emus and bustards carry over great distances in the desert night air, and those of bitterns effectively penetrate dense reed beds.

Meanwhile, other non-passerine groups have evolved their own sonic strategies. As we've heard, cuckoos use clear and far-carrying whistles to announce their presence across the landscape. Waterfowl often have expressive voices to communicate with their own kind (and others) among the communality of the wetlands. Clans of kookaburras bond by calling in chorus with loud voices, often in the predawn and dusk twilight.

One possible sonic strategy that really intrigues me may occur in the nightjar family. These fully nocturnal birds are aerial hunters for insects, sweeping back and forth across open areas on sharply pointed wings like huge swallows. They'll often call at night with vocalisations distinctive of their species. 🐦 However, both in the Australian outback 🐦 and the teak forests of India, 🐦 I've heard nightjars making another type of sound. It is highly un-bird-like; a series of loud twanging or clicking vocalisations. They are such short, transient pings that I wonder if they function as echolocation while hunting. If so, then nightjars would join only a handful of other birds, including cave-roosting oilbirds (a close relative) and swiftlets, in employing such a sonic strategy.

Throughout the morning at Lore Lindu, I continued exploring the magnificent primary rainforest, encountering species endemic to Sulawesi. A group of Piping Crows moved through the treetops, delighting me with their sing-song musical cries – so unusual for a corvid. ♪ The warm-toned whooping of Bay Coucals drifted up from the undergrowth of ferny gullies. ♪ Sunlight shafted through the humid air, and at one point a pair of Knobbed Hornbills aero-braked into the canopy with whistling wings, honking noisily as they commenced feeding on figs. ♪

As I was musing on the wonders of the place, a sound of delicate tinkling materialised into a caravan of gaily-decorated donkeys coming down the path. With bells jingling around their necks, they were laden with produce for market and accompanied by a group of young villagers, to whom my presence seemed to cause great amusement. ♪ When Sarah caught up with me shortly after, she said that they'd also paused to talk to her, and were apparently most concerned to find us alone in the rainforest. It was difficult to explain in our hesitant Indonesian how special it felt for us to be there.

Reluctant to head back to the bustle that would be gathering around Ira's homegrown cafe, we hung around just enjoying the rainforest. Butterflies were everywhere, floating by in all colours imaginable. There were green ones and yellow ones, large ones and tiny ones, black ones with swallow tails and iridescent blue ones. Occasionally, one would alight on our outstretched hands.

I knew that the renowned nineteenth-century naturalist Alfred Russel Wallace had collected butterflies in Sulawesi, being fascinated by their diversity here. His sojourn in Sulawesi informed his ideas, and his correspondence with Charles Darwin helped shape the theory of evolution, of which Wallace is now acknowledged as co-originator.[39]

So here we were, in the cradle of evolutionary ideas, having heard a baritone descendant of ancestral pigeons, with Wallace's butterflies landing on us to sip at our sweat.

When Sonic Strategies Fail – Regent Honeyeaters

If a sonic strategy can facilitate survival, can listening also give us a clue as to why a species is struggling, and inform conservation efforts? Australia's Regent Honeyeater may offer a story.

Regent Honeyeaters are handsome birds; black with gold edges to wing feathers and chevrons across the back. They are a large honeyeater, most closely related to the wattlebirds. Like them, they feed on nectar, but are more specialised, preferring a handful of eucalypt species that flower profusely with high nectar production. Being a larger nectarivore, they require a high-energy diet to maintain activity and breed successfully.

Previously, the inland slopes of the Great Dividing Range of Australia's southeast provided a vast resource of fertile woodlands and mature flowering trees. Historically, regents were recorded ranging widely across the region, as far as Adelaide in the southwest and north well into Queensland. Reports from early settlement describe them as common, with sizeable aggregations appearing when eucalypts were in flower. Flocks of a hundred or more were noted, and they were often seen flying down the streets of regional towns. They would once have occurred in the box woodlands where we live, and perhaps even the old trees on our land would have known visits from roaming bands of regents.

But no more. As broadacre farming cleared extensive areas of their ancestral habitat, the range of Regent Honeyeaters contracted dramatically. With the most fertile areas turned over to agriculture, and remaining forest prioritised for timber harvesting, their numbers have dwindled alarmingly. There may be only a few hundred Regent Honeyeaters left in the wild, in scattered populations dependent on a few remnant patches of suitable habitat. They have been listed as critically endangered – as dire as it can get. Now, the nearest regents to where I live are found several hundred kilometres distant, at Chiltern National Park. Although a reliable population, their numbers at that location are possibly only a few dozen.

In 2019, I helped convene a national conference for wildlife sound recordists on the NSW central coast. One of the speakers invited by my colleague Sue Gould was Mick Roderick, the Woodland Birds co-ordinator for BirdLife Australia. Arriving as planned around lunchtime on the first day, he looked distracted. Greeting him, I asked if everything was OK. He replied that, as he'd arrived a little early, he'd taken a few minutes to explore down a dirt road nearby, and had just come across a group of Regent Honeyeaters feeding in Swamp Mahogany. It seemed there may be half a dozen birds there. His excitement was due to this being the biggest group of regents that had been found so far that year.

Needless to say, our speaking schedule was sidelined as my colleagues set off with microphones and cameras to document our rare guests. When I went down later, I found the birds still there, as Mick had predicted they would be, feeding discreetly among the stand of gums.

For a large honeyeater, regents have a gentle demeanour. Wattlebirds and friarbirds, Australia's other large honeyeater species, are garrulous and noisy, their loud calls announcing their presence throughout the day. In contrast, I found it a testament to Mick's birdwatching skills that he'd noticed the regents at all. They were rarely calling, instead moving inconspicuously through the foliage, only vocalising occasionally. When one does hear it, the song of the Regent Honeyeater is quiet and lyrical. With sweet and tender phrases, often punctuated with bill clicks, it is a song that reflects the bird's personality. 🐦

And this speaks of the regent's problem. The clearing of land has been cataclysmic enough for them. But along with it has come an increase in the numbers of more generalist honeyeaters such as wattlebirds, friarbirds and miners. They have multiplied, adapting to new opportunities such as ornamental gum trees planted in towns and open spaces. The regents don't have the bold temperament for this, they prefer the quiet life. It is also why Mick was confident the birds would remain feeding in the Swamp Mahogany; they'd found an unobtrusive place to themselves where they weren't being disturbed by their rowdy cousins.

During our conference, David Stewart hosted a discussion on mimicry. He noted that the honeyeater family, numbering nearly two hundred species, around half of which are native to Australia, are not known to mimic. With one exception – the Regent Honeyeater. And the species they imitate? Wattlebirds and friarbirds. My speculation is that regents do so as a form of acoustic distraction, pretending to be a more extrovert species to avoid being hassled by them. If so, this is a remarkable sonic strategy. However it becomes more intriguing when you consider regents would likely have had less need for it before their habitat was cleared and other large honeyeaters became numerous. Is it an ancestral strategy to deal with their boisterous relations? Or a very recent change in response to their current pressures – vocal learning at its most adaptive?

In light of this idea, something else that had been puzzling me fell into place. Chiltern is a remnant area of ironbark woodlands. It is outstanding habitat, however a national highway runs right through the middle of the park. Recording there is frustrated by traffic roar that is pervasive and continual – you can't get away from it. Which may be just the way the regents like it. The site offers them both rich nectar opportunities and 24/7 sonic camouflage. In the breeding season this would be a liability, but as a wintering location, which is when they're there, the acoustic smoke screen of the highway may be advantageous.

By considering these aspects of a species' sonic life, we may gain a deeper insight into their requirements, and how best to focus our efforts in assisting them. Conservationists know that Regent Honeyeaters require the restoration of suitably rich habitat. But regents may be telling us that they're also introverts, and need refuges from being hassled by other large honeyeaters.

Recently it has been suggested by researchers that the mimicry of other honeyeaters is a sign that regents are losing their own song. I hope that my interpretation is closer to the mark, because if gaining a song marks the emergence of a new species, then losing one must be a signal of its imminent and terminal decline.

It would be a tragedy to lose these beautiful birds. I hope that the sweet songs of Regent Honeyeaters, far from being lost to our world, may instead contribute to saving them.

Chapter 8

The Mind of Nature

The Sonic Cycle of the Cloudforest

I was gently rocked awake by an earthquake. It was probably only a tremor, but lying on the ground in a tent and amplified by my swaying air mattress, it felt quite strong. I recalled my father's familiarity with earthquakes, having lived for a time in New Guinea. On the occasion during my childhood of a rare quake being felt in Sydney in the small hours, he responded to my mother's alarm at the house shaking by mumbling for her to "go back to sleep, it's only an earthquake."

Now it was the small hours of the morning, and I was in New Guinea, in a bush camp in the highlands with my sound recording colleagues and our local guides. No sooner had I thought to roll over myself, than a tearing, groaning and shredding signalled a tree destabilised and collapsing. The swish of foliage and vines torn asunder became a final roar followed by a great thudding blow as its bulk landed frighteningly close by.

Voices were raised and flashlights lit, as our guides leapt to find what had happened. I heard the hacking and slashing of bush knives as they cleared debris, however they didn't seem too concerned, so I knew we'd been fortunate. I lay back, tired. After nearly a month in the highlands, trekking muddy paths through a primal landscape, recording at every opportunity, I was exhausted. It was one of our last days in the cloudforest, but all I wanted to do was put my head under a pillow – which I didn't have, just a rolled up jacket.

The heavy rain of the previous evening had stopped, and realising the morning was not an opportunity to be squandered,

I wearily rose and prepared for the jungle. This involved applying liberal amounts of insect repellent on socks to stop tiny mites from turning my ankles and feet into red, inflamed horrible things, plus more on face, neck and hands to deter mosquitoes.

But the quake and treefall had rattled me, and this morning I didn't feel like setting off alone in the dark as I usually did. I asked one of our guides, George, now nonchalantly fanning our ailing cooking fire back into life, to accompany me. After a quick cup of tea, he grabbed his bushknife (without which highland men rarely go anywhere) and flat cloth cap (which wouldn't have been out of place in an English village), and we set off.

The tree had indeed fallen right on the edge of our camp, missing it by a mere few metres. George chopped a few remaining branches clear, and we pushed our way through the tangle. The location I had in mind to record was one I'd scouted out the day before. It was about a kilometre distant, accessible via faint footpaths through the rainforest. I was confident I could navigate there myself, but it felt reassuring to have George with me.

The place I'd chosen was just off the side of a ridge-crest. In this landscape, it was an unusually level basin with tall, mature trees emerging from dense undergrowth. I hoped this natural amphitheatre would provide the focus of a rich soundscape. This was the first trip where I was employing a new microphone setup, an 'iso-binaural' rig I'd designed to capture sounds equally from all directions. All I had to do was situate it in a 'sweet spot' where activity would be concentrated, and the microphone configuration would capture a richly spatial, stereo recording. Setting up my equipment, George and I retreated some distance to listen.

Once headtorches had been switched off, the sound of the predawn forest seemed even more overwhelming. A million unseen insects chimed, trilled, zizzed and chirruped in a pulsing cloak of sound. Mixed in with this were the penetrating peeps of tiny frogs that seemed to be emanating from everywhere on the forest floor.

This nocturnal chorus had been filling the tropical air all night, but soon the birds were waking. Jewel-babblers, Regent Whistlers,

Friendly Fantails (yes, their species name), Greater and Lesser Ground-robins, Ornate Fruit-doves and Papuan Mountain-pigeons were among voices that began weaving a dawn chorus. Over half an hour it developed, with many of the species possessing tuneful songs, together forming a pleasant mingling of pitches and tones. It was one of the most melodious dawn choruses I have ever encountered. As the light grew, George pointed out a group of secretive Spotted Jewel-babblers moving around among fallen timber nearby while giving limpid, piping calls that seemed to counterpoint all the other voices.

As the dawn birdsong reached full abandon, and I could not imagine anything outdoing them, the cicadas started up. This was a regular daily occurrence. Their effervescent fizzing quickly became so loud as to obliterate everything else. They could be deafening in proximity – there were occasions during our trip when I had to move away if one began calling on a tree close by. Despite them, the birds continued undaunted.

The whole forest felt so gloriously and riotously alive. 🐦

An hour later, the cicadas had stopped and the birds subsided slightly. Feeling somewhat more invigorated, I suggested to George that he may like to head back to camp for some breakfast while I remained. As the morning progressed, far from waning, the birdsong picked up again. 🐦 I wandered around photographing for several hours, before hunger got the better of me and I too walked back, leaving the microphones running in situ.

When I returned well after midday, everything had quietened down considerably. I crouched by my recorder, still faithfully operating. Not that there was much to record. In contrast to the dawn when the volume level meters had been lighting up brightly and continually, now they were hardly registering anything at all. I closed my eyes and listened. Almost nothing – even those tiny frogs had gone silent. It was a dramatic contrast, from over-whelming to nary a whisper. The thought occurred to me that it had now gone unnaturally quiet. 🐦

Yet there was nothing unnatural about it. After weeks in the highlands, I was completely familiar with this afternoon stillness.

It happened every day. I knew it would continue until, as dusk approached, there would come a moment when a single cicada would herald a dusk cicada chorus. Within seconds the air would shimmer with the calls of what my companions had come to describe as 'dentist drill' cicadas. These were a different species to the dawn cicadas; shriller and, if it were possible, even louder. For twenty minutes or so, you couldn't hear yourself think, as great waves of shearing, screaming sound tore at the air. With darkness and the last stuttering cicada, a great choir of nocturnal insects would be revealed as having taken over. And so it would continue throughout the night, until the next morning brought a new dawn chorus.

This was the sonic cycle of the cloudforest, the audible patterning of nature, as unchanging as time itself. This diurnal cycle was something I was familiar with from tropical forests elsewhere – the western Pacific, southeast Asia, India – but here, I was drawn to puzzling about it. The contrast between afternoon hush and other times of day had piqued my curiosity.

My first thought was that it was a response to the heat of the day; nature's time for a nap. For us, siestas are an adaptation to noonday heat and a cultural life oriented to cooler evenings. However in tropical forests, air temperatures remain reasonably stable throughout the day and night. While often drowsy in the afternoons myself, I knew this was simply the sleep deprivation of a nature recordist who'd been up half the night.

Another explanation was that insects and frogs are more active at night, birds during the day, while cicadas are crepuscular (during twilight, at daybreak or dusk), at least acoustically. This cycle of activity would result in a natural lull in the afternoon.

Yet immersed in the highland cloudforest as I had been these weeks, my instinct suggested there was more to it than simply an artefact of activity. I sensed something in the stillness; a presence, an intention. I heard this mid-afternoon hush as an intrinsic feature of the ecosystem, with its own significance.

Acoustic Biodiversity

Sound in nature results from activity and tells us what is happening. On an ecosystem level, we should thus be able to hear whether a habitat is healthy – whether all systems are functioning well. Of course only a subset of creatures in the whole ecosystem will be soniferous, yet all life is interdependent, so that should be enough to indicate wellbeing throughout.

But what, specifically, should we be listening for? What aspect of the soundscape will point to ecosystem health? You'd expect the answer to this would be obvious, but actually, it is something of a puzzle. During the last decade, ecologists have been embracing sound as a way of monitoring and assessing ecosystems worldwide. This accelerating enquiry has given rise to a new research discipline, ecoacoustics,[40] built on a previous one, acoustic ecology, which originated during the 1970s from more humanities-based studies.[41]

Whatever one calls this new enquiry, science is now having its 'Mutawintji moment' of acoustic awakening. A myriad of passive acoustic monitoring devices are being deployed in wild locations. From these robust and programmable audio recorders, ecologists are now collecting a wealth of data. Analysis is assisting in the recognition of many rare species in places where their presence was uncertain, and allowing researchers to reliably catalogue the biodiversity of remote habitats. It is a significant advance, and one that speaks of the capacity of sound to offer us important information about the living world.

When Associations are Lost

However, this focus on biodiversity is only part of the tapestry of information required to assess the whole ecological balance. Why would biodiversity alone not be a measure of an ecosystem's condition? Let me give you two personal experiences that may offer some perspective. They're both somewhat extreme, but nevertheless demonstrate that there is far more to an ecosystem than simply a census of what lives there.

Prior to flying to Papua, I'd spent a few days with friends in Brisbane. They live in a leafy suburb with a bush park nearby and tall trees dwarfing their house. It is good natural habitat, and like the suburbs of many Australian cities, birds have established themselves well. I awoke to a chorus of magpies, butcherbirds, lorikeets, miners, wattlebirds and kookaburras. These are all large birds. What I was listening for were the smaller ones that should have been present, but weren't: fairy-wrens, robins, thornbills and scrubwrens.

So listening alone told me that the species diversity was awry. This, sadly, was no surprise. Small birds have been steadily disappearing from our urban areas around the country. Predation is almost certainly implicated, partly from introduced cats and rodents, but also from those larger native bird species. Which prompts the question; why is their predation significant in the suburbs when small birds co-habit with these same larger species in wild locations? Whatever the specific answer may be, it basically points to the suburbs as being a modified ecosystem. The creatures that live there are either previous species still hanging on, or adaptable newcomers able to exploit a new situation, which all those larger birds are.

But there was more to what I was hearing than simply this species imbalance. The whole place sounded out of whack. The magpies, ever expressive as they called from the neighbour's washing line, had a shriller repertoire than those I'm familiar with in the bush at home. The butcherbirds too were more strident than any I'd encountered in the wild. Everything sounded edgier, more insistent. Even the lorikeets sounded a bit demented.

But the thing I noticed was that throughout the day, this background of agitated birdsong continued. There was no quiet lull in the afternoon. One could say that the place had lost its rhythm.

*

Years previously, Sarah and I had spent several months on a field trip to southeast Asia, during which I'd recorded in some of the most ancient rainforests on the planet. These ecosystems in Thailand and Malaysia – Khao Yai, Kaeng Krachan, Khao Sok – have been relatively stable environments over many tens of millions of years. Immersed in a lush green world, my ear settled into their cycles of activity and quiet.

In the last days of the trip, having emerged from the jungle, I visited a renowned bird park in a major city. I'd normally avoid zoos, but I was maybe hoping for a few photographs of rarer tropical species. Within minutes of entering the front gate, I knew I'd made a mistake. Following a path, I came upon a group of visitors jostling around something in their midst. Approaching I saw a solitary penguin had been taken from its enclosure and was being paraded in tropical heat so that visitors, grabbing at its flippers, could take selfies. The poor creature looked terrified.

Feeling distressed myself, I turned and hastened on to the tropical aviary. It was a netted dome covering a few acres, with a double airlock style entrance, presumably so that any birds would be unlikely to escape. Yet there were no free flying birds inside. Instead they were all confined to smaller enclosures, many species packed in together. I spotted Nicobar Pigeons, Fire-tufted Barbets, Pied Hornbills, trogons, and birds of paradise. They even had a few critically endangered Bali Mynas.

In the wild, I'd be thrilled to encounter any of these exotic species. But here, I was feeling further anguish. It was partly the sight of a Bali Myna having pulled out its feathers due to stress, and a rail, normally an extremely shy and retiring species, sitting limply out in the open, its eyes half closed. But mostly, it was the sound. I cannot describe the bedlam of screeches, cries and squawks that besieged me. I was dismayed hearing the agitation of those birds, their distress overwhelming and continual, coming from all sides. After months of listening to ancient rainforest ecosystems, the contrast was absolute.

Once again, here was diversity of a kind, but no healthy environment. The birds were vocal, but they told of misery, of a place where the natural patterns of their existence had been stripped away.

The Quiet Time of Day

So, in the afternoon stillness in the Papuan cloudforest, I found myself thinking about ecosystem integrity. This train of thought drifted into reflecting on our human ideas of health. Many cultures recognise health as embodying a whole-of-life integration. Various contemplative traditions focus on achieving this through meditation practices, or creative and artistic activities. Even exercise can be equal parts exertion and calm alertness. There are factors common to all these activities which seem to generate wellbeing. They require a focused state of awareness, a form of stillness, a turning of one's attention inward rather than on external stimulation, and they are most beneficial when practiced regularly – daily is often ideal.

It occurred to me that each afternoon the entire cloudforest was doing something similar. This may appear an unlikely idea, that an ecosystem could be 'meditating', but bear with me, and let's explore the possibility.

The most significant consequence of an afternoon hush is likely to be cognitive. When creatures are being communicative, they will be attentive to signals from both their own and other species. Some of these signals need to be responded to, and others ignored, but all need to be cognitively processed. In whichever way animals do this, it will form a substantial component of their mental activity.

When the forest goes quiet each day, an outcome has to be that creatures have less overall information to process. Not only is their own communicative activity minimised, but so is their need to process the signals of others. As a result, stimulation levels across the whole ecosystem decrease.

If the afternoon hush functions to regulate cognitive activity, then it could perform a similar biological role for other sentient creatures that calm awareness does for us. Listening to the whole habitat, patterns of daily sound and silence can be heard as speaking of the relationships which maintain and regulate the integrity and functioning – the health – of an ecosystem. Clear and distinct patterns can be interpreted as a feature of mature ecosystems in which organisms have adapted to living together over immeasurable periods of time.

These daily sonic cycles are recognisable in ecosystems around the world. They're spectacularly noticeable in tropical regions, due I think to a combination of reasonably stable climatic conditions (outside of monsoons, at least) and the vibrancy of the soundscape when it's 'on'. As one moves in latitude away from the tropics, seasonal factors have more marked impacts on the natural soundscape, yet daily cycles remain recognisable despite more changeable weather and the seasonality of breeding.

These cycles are also evident in their absence. In relatively immature ecosystems such as the Brisbane suburbs, a community of newly establishing species are still working out how they're going to live together, and clear patterns have not yet emerged. In dramatically disturbed ecosystems or artificial situations such as zoos, a functional ecology may not be possible. In these cases, the agitation of voices and absence of any clear patterns are very noticeable.

That daily contrasts of sound and silence in a habitat could be indicative of deeper levels of ecosystem organisation is intriguing, but what might the process underlying such a phenomenon be?

Homeostasis

As ecosystems form and mature, they go through a succession. From a community of establishment species, a range of additional species will gradually settle and integrate. In time, the community of organisms becomes biodiverse. So in terms of sound, assessing acoustic diversity is a reasonable first measure of the maturity of an ecosystem.

With this species diversity comes an increasing complexity and number of interactions between those species. Levels of synchronisation and co-operation emerge. In ecosystems that have grown undisturbed over aeons, these associations form webs of relationship on which everything depends. In this way, living systems can be described as self-generating and self-organising.

Ecologists are seeking to understand the processes that result in such self-regulatory outcomes. Many take the form of feedback loops which govern the interactions of organisms and their environment. They can be understood as being of two kinds; positive and negative. Taking ourselves as a self-regulating organism, hunger is a familiar example of a negative feedback loop. A perception of hunger triggers us to eat, which then negatively influences appetite, until waning energy levels once again increase hunger. Conversely, addiction is a positive feedback loop in which indulgence often leads to more desire and less satiation. So while negative feedback loops tend to result in stability and homeostasis, such as maintaining body nutrients at optimal levels, positive loops amplify processes and may lead to a runaway effect.

While negative feedback loops can be expected to play a key role in generating the continuity we observe in living systems, positive feedback loops may also have their place. I suspect that the afternoon hush of the Papuan cloudforest is the result of one of these positively reinforcing processes, the quietening of each creature having the effect of quietening others, until the forest falls largely silent.

As an isolated process, such a positive feedback loop could be detrimental, and in this instance we could imagine it leading to a perennially soundless landscape. But of course, it exists within a complex biological system. For positive feedback loops to fulfil their function rather than become runaway, rogue processes, there needs to be a disruptor, a circuit breaker.

In tropical forests, the screaming dusk chorus of cicadas provides a reset on a daily basis, and in no uncertain manner. As daylight fades, the air is charged with sound of extreme volume. By the time the cicadas recede back into silence, the whole acoustic system has been triggered into a new state, one in which nocturnal insects chime, click and buzz on throughout the night. Thus a combination of sonic feedback loops could lead to both spectacular daily contrasts and long-term homeostasis.

When I hear clearly defined temporal cycles, I can be confident I'm in an ecosystem that is stable, mature, healthy in its function and regulating the circumstances of its own wellbeing. This systemic symbiosis is so important that, unlike we who have to choose to 'decompress', living systems have it as default. I felt that in the afternoon stillness of the Papuan cloudforest, I was attuning to a synchronicity among all the organisms of the place, a whole-of-life integration. In addition to hearing the sentience of individual creatures, I was listening to the sentience of the ecosystem – the mind of nature.

Wouldn't it be ironic – given that a heartbreaking silence accompanies biodiversity loss – if the measure of health in natural systems were to be found in listening for a more dynamic expression of quietness?

Chapter 9

Avian Co-operation – Birdwaves

In Sulawesi's Cloudforest

A 'birdwave' is a phenomenon of the natural world. They are a spectacle, and never cease to inspire in me a childlike amazement.

The first time I knowingly encountered one was in the highlands of Sulawesi. At 2000 metres, every tree on the summit of Gunung Rano Rano was adorned with epiphytic mosses. Stunted and twisted, they were clothed in soft green and hung with copious drapes of growth. If I leaned a hand against a trunk, it could disappear up to my wrist in the sponge. The place was sodden, both arboreally and terrestrially, with every footfall sinking gently into a pliant carpet of moss and detritus. This made the yielding ground unpredictable, and I had to step carefully to maintain my balance. Water dripped everywhere, and fogs of mist could sweep in at any time.

After weeks in the humid lowland rainforests of Sulawesi, to be in this cool cloudforest was somewhat of a relief for Sarah and I. Our local guide and his two friends however, acclimatised to heat, were shivering as they huddled over a smoky fire. Young and aspiring to be tourist guides, we'd christened them 'the likely lads', as their professionalism had a way to go. Our guide had learned his English listening to pop songs, which gave him a curious way with words. He'd already nearly got us lost ascending the mountain, the 'chicken' in the menu plan had turned out to be cracker biscuits, and the previous evening they'd improvised a meal for five using a single tin of sardines mixed with boiled fern shoots scavenged from around the camp. Nevertheless, we were delighted to be here, and they seemed happy amusing themselves telling stories.

The night had been alive with the rattling calls of small tree frogs, a chocolate brown species I'd found clinging to shrubs and fernery. 🐦 With the dawn approaching, they had quietened, and it had begun raining, a gentle precipitation that fell imperceptibly on the absorbent forest floor. As the dawn chorus grew I was recording species endemic to the island; Maroon-backed Whistlers, a Rufous-sided Thrush, Mountain White-eyes and flycatchers. 🐦

After the dawn chorus waned, I wandered around, watching and listening for further activity. It had fallen very quiet. A lone Crimson-crowned Flowerpecker flitted into a bush, its brilliant red and black plumage reminding me of our Mistletoebirds at home, to which they're related. Hearing a prominent call, I followed it and found it uttered by a Spot-tailed Goshawk, a surprise as it seemed an unusual sound for a raptor. 🐦 But then, Sulawesi was turning out to be full of surprises.

A little later, out of the corner of my eye, I noticed a single bird fly in and alight in a nearby tree. In the few moments it took me to get my recorder running, a few others had flown in to join it. They hopped along and under branches, inspecting tirelessly among epiphytes and mosses, pecking excitedly and all the while keeping up thin, whistled contact calls.

Meanwhile a commotion of other sounds was approaching, and suddenly it seemed birds were coming from everywhere, among them Leaf-warblers and White-eyes. A drongo, unmistakable with its deeply-forked fish tail, alighted overhead, while a Citrine Flycatcher appeared in a bush nearby. A family group of Malias came through the understorey, bouncing along the ground and into low shrubs, adding lovely growls to a growing cacophony of trilling, twittering and chipping. 🐦

The place had become alive with movement and sound. Everywhere I looked there seemed to be birds; climbing up tree trunks, investigating the foliage, perching while scanning the ground or flitting through the undergrowth. Every leaf seemed to have a bird hanging from it. Their calls rained on my ears like an audible monsoon.

For the next ten minutes, it was activity central. And then this mass of birds, all traveling together, began to move on, and I sensed the bustle around me subsiding. In groups they flew off, until the last few followed the others with a purr of wingbeats. Silence again. Like they'd never been.

I'd just experienced a mixed species foraging flock. I estimated there must have been over a dozen species involved, and nearly a hundred individuals. For a birdwave, it was a small one.

In India's Broadleaf Forests

Six months later, we were in India. For me, it was my first time back after travelled there extensively in my twenties, but for Sarah, her first experience of the country.

When we mentioned to friends that we were going to India to record nature sound, we were often met with polite incredulity; 'why would you want to go to the most overpopulated country on the planet to find nature?' Descending into Mumbai airport over vast expanses of slum dwellings, we asked ourselves the same question.

Yet India is a place of polarising contrasts. Yes, it's crowded, noisy and polluted, but the people are welcoming, worldly and open. Their rich cultural history stretches back millennia, and the sheer colour and vitality of the place can take your breath away. One minute you'll be inhaling the scents of jasmine, incense and spices, the next half retching as you walk past a noxious pile of refuse.

Away from the cities, India has a well-maintained network of national parks. These areas protect a wealth of the subcontinent's wildlife: elephants, bison-like gaur, a variety of deer, peacocks, giant squirrels, monkeys, sun bear, a dazzling diversity of birdlife, and Shere Khan, the Bengal Tiger. It is a nature recordist's paradise.

Or so I was anticipating as I set out for my first morning of recording near a seaside village in Goa, which we'd chosen as a peaceful place to rest and acclimatise before beginning our field work in earnest. Surrounded by fields and a mosaic of forest, it wasn't exactly wilderness, but it was India, and I was looking forward to what I'd encounter.

Setting up in the darkness predawn, the first thing I heard was the soft cooing of a Spotted Dove. In Australia these are an introduced species, common in most urban areas around the country. Then came the pleasant tinkling song of a Red-whiskered Bulbul. I recognised it immediately as it's also an import – I'd had them breeding in a camellia bush outside my childhood bedroom in Sydney. 🐦 Next was the scolding chatter of Common Mynas, another ubiquitous, introduced bird from my youth. Then the House Sparrows woke up.

It was all a little deflating. However I continued exploring the forests beyond the edge of the village that morning, eventually hearing my first flameback, an impressively-sized woodpecker, and witnessing langur monkeys crashing excitedly through the treetops above me.

It was only years afterward, when I came to listen again to those recordings with more experience, that I found what the langurs were so agitated about. While standing alone in the scrub with my microphones, I'd captured the low growls of a Leopard. They don't call loudly – from the recording I estimate it was probably only a hundred metres away. 🐦

Later that week, we were settled at Cotigaon National Park in the south of the state, exploring a vast region of broadleaf forest that backed up into the hills of the Western Ghats. The motor scooters we'd hired seemed to have no suspension to speak of, which made our daily forays along the stony roads into the park somewhat uncomfortable, but once there, we were in a magical realm.

As dew fell from the treetops before dawn, tiny Jungle Owlets yapped in the dark before roosting, and the first Racket-tailed Drongos began calling. They were followed by the song of one of the most melodic songbirds on Earth; the Malabar Whistling Thrush. They sing a slow and somewhat hesitant melody at the first hint of daylight, a tune drifting absent-mindedly through the forest. It really is one of the most beautiful of birdsongs, and I find it puzzling that it is not more celebrated. 🐦

I must admit that at this point, I'd forgotten about mixed species foraging flocks. Perhaps I hadn't registered just how characteristic of warm latitude forests they could be. Later that morning, as Cotigaon forest was transformed by a first few birds flying in, materialising as if from nowhere, my excitement was rekindled.

Soon, I had a swarm of wings, feathers and twittering voices around me. Tiny Velvet-fronted Nuthatches inspected the bark of branches and tree trunks, often hanging inverted to do so. Yellow-browed Bulbuls chattered amiably and Orange Minivets flitted among the foliage with silvery whistles. Flowerpeckers and warblers moved with them, adding chipping contact calls. There were leafbirds, a pair of Indian Paradise Flycatchers, Asian Fairy-bluebirds, and sunbirds which skittered around the mid-storey. I even caught sight of a handsome Malabar Trogon hiding among the foliage. Nearby, a Brown-breasted Flycatcher moved from one perch to another, scanning the forest floor, while a squad of Brown-cheeked Fulvettas gave penetrating, sing-song calls as they patrolled through the undergrowth. From the crowns of the trees to the forest floor, birds were everywhere and in motion. Among them all, Racket-tailed Drongos flew out in aerial pursuit of insects. 🐦

It turned out that this was also a relatively small party. On other occasions in India we were to encounter flocks numbering hundreds of birds of perhaps several dozen species. One such group, at Thattekad Sanctuary in Kerala, was so numerous as to take nearly an hour to pass us by and move off through the forest.

The Pied Pipers

Our Indian naturalist friends (who call themselves 'wildlifers' rather than 'birders' – why wouldn't you when you have tigers to observe?) were the ones who introduced us to the term 'birdwave'.

They also told me about the Racket-tailed Drongo, those birds I'd noticed hawking for insects in the midst of the flock. They're boldly-natured and striking birds; iridescent black with a pair of long tail streamers ending in flared rackets. As with other drongos, racket-tails are accomplished mimics, capable of imitating dozens of species with whom they share the forest. Our friends told us of their suspicion that, through its mimicry, the Racket-tailed Drongo was drawing the whole flock together and leading it through the forest.

A few years later, a paper published by a group of American and Sri Lankan researchers suggested they were correct.[42] Mixed species foraging flocks are, as I'd understood, composed mostly of insectivorous species – those that glean insects from foliage, from the canopy to the understory. Ganging up together on insects is a successful behaviour for birds, as the collective commotion disturbs and flushes them out, making foraging easier and more successful for everyone. For drongos, it has been measured to increase foraging efficiency by nearly 50%. Some observers have suggested that drongos rarely forage alone or outside of associating with flocks. They will even adjust their height in the forest strata to remain in proximity with other foraging birds. All this implies that drongos are quite dependent upon feeding flocks.

Among the species that frequently join foraging flocks, some are considered 'nuclear'; the core members around which the flock aggregates. In India and other Asian tropical forests, various drongo species seem to fulfil that role.

The Racket-tailed Drongo's mimicry is so accurate that it has been shown to attract other birds to them. Thus by imitating, the drongo is implicated in drawing the feeding flock around it. Like a bugler, it calls the hunting party together. It can be speculated that this is an adaptive behaviour, a sonic strategy that has been pursued to the obvious benefit of the drongo. But what about the other species? What do they get out of this arrangement?

Drongos are aerial feeders, sallying out in aerobatic sorties to snap up insects on the wing. They're sharp-witted, keen-eyed and vigilant, their flights allowing them to scan for predators such as Accipiter hawks like the Besra. At first sight of danger, the drongo's loud and incisive alarm calls give other birds an unmistakeable warning. Drongos thus fulfil the role of sentinels, allowing other species to focus, heads down and tails up, on scanning bark, leaves and undergrowth for insects. On very rare occasions, a bold drongo may pinch someone else's meal, but it's a small price to pay for hanging out with the prince of early warnings.

Listening back to my birdwave recording from Cotigaon, I found that as the flock approached, a Racket-tailed Drongo calls out, sharp and clear. The same is heard on my Thattekad recordings, and several others. Whether I've also captured these birds imitating I can't tell, as they would blend undetectably with their mimicked species, but it is affirmative to hear that they were present on each occasion.

A Global Phenomenon

Birdwaves are not only a feature of Asia's tropical forests. They are apparently spectacular in the forests of central and tropical south America. Nuclear species in this region have been observed to include various members of the antbird, warbler and tanager groups, which often have striking plumage and vocalisations that attract other species.

They are also found in more temperate latitudes, and we have our own birdwaves in the bushland at home. Flocks here are smaller, only involving a handful of species, and usually encountered in the cooler months when breeding has concluded. Thornbills such as our Buff-rumped or Brown species are core members, often joined by Weebills, treecreepers, robins, sittellas, whistlers and fantails. Interestingly, while fantails are aerial feeders, the thornbills are notable mimics. Who does everyone follow? The mimics. Thornbills are also vigilant, and likely act as good sentinels. 🐦

Foraging flocks similarly occur in northern temperate regions, once again in the non-breeding season, where they may be led by tits, chickadees and nuthatches. Like our thornbills and Weebills, these small birds give constant contact calls which likely support cohesion of the whole group. 🐦

Mixed species foraging flocks thus bring together a great diversity of insectivorous songbirds (plus the occasional non-passerine such as my party-crashing Trogon) into a co-operative way of living.

As such, these associations highlight the significance of sound, and in particular song learning. It has not only resulted in songbirds finding new ways of communicating with their own kind, but opened the possibility of this degree of inter-species co-operation, as mimicry seems integral to the whole exercise. Indeed foraging flocks could be the reason birds such as drongos mimic so accurately – they've come to do so because it has a positive benefit on their foraging success. One can imagine an evolutionary process by which vocal ability, feeding efficiency and species association become mutually reinforcing, resulting in the complex web of relationships we encounter in flocks today.

Whenever I've returned to Asian forests, and find myself waiting in the quiet of the jungle for a birdwave to appear, I muse on those birds that will arrive. I imagine them waking at dawn, having a stretch, perhaps a little sing, and then heading off to join the flock. In the same way that we might go to a cafe for breakfast, they'll fly off to find a drongo. For the rest of the day they'll roam the forest, following their pied piper, feeding and hanging out in the safety of commensurate species.

The web of mutually beneficial relationships involved in this co-operative behaviour is probably unfathomable. It involves not just a handful, but often many species. Some may be core members, while others only join the flock for a short while. Each species will have their own ways of foraging, yet they come together as a community to do so. In this way, they may spend much of their days, and ultimately lives, together. All around the planet, from the tropics to the high latitudes, different aggregations of insectivorous birds unite in this behaviour. It synchronises their most essential daily activities; feeding well and remaining safe.

This profound degree of co-operation is all made possible by acoustic behaviours. Where else might this depth of co-operation in nature be audible? Among birds, an obvious possibility is the dawn chorus.

Chapter 10

Avian Diplomacy –
The Dawn Chorus

A common interpretation of the dawn chorus is that birds are singing to compete for mates and defend their territory. I must admit that in all the years I have been rising early to record dawn choruses around the world, it would never have occurred to me that I was hearing something essentially competitive. Because I don't believe it is.

Quite the opposite. I hear the dawn chorus as one of nature's most sophisticated processes of negotiation – both between species, and among members of the same species.

Like foraging flocks, the dawn chorus is a worldwide avian phenomenon. It is to be heard from sub-arctic and temperate regions through to the tropics. In terrestrial habitats, from the first hint of daylight or sometimes well before, birds will begin filling the air with song. It will be familiar to anyone fortunate enough to have them roosting nearby and be awake at that hour to listen in.

Why are these choruses so characteristic of the hour before sunrise? A combination of factors seem responsible. As first light pales the horizon, birds awake. Twilight is not particularly suitable for foraging, yet it does afford them protection from predators. As there is vulnerability for a bird advertising its location through extended singing, predawn is a good time for doing so, because they will be less visible.

Then there are the atmospheric conditions. After the cool of the night, the temperature is usually at the lowest it will reach over a 24-hour period. This is when the transmission of sound waves over distance is greatest, and birds can be heard the furthest.

In conditions of low humidity, such as desert locations or during dry weather, this dawn temperature effect can be pronounced, and birdsong audible from far off across the landscape. In the higher humidity and more uniform temperatures of tropical regions, the effect may not be so noticeable, but is still present. So it is generally around dawn that the acoustic horizon is the most extended.

It is also when the air is most likely to be still and calm. Unstable weather, such as my blustery morning at Kata Tjuta, will put a dampener on birdsong. Gusty wind conditions limit sound transmission, and will also give rise to noise through foliage. Notwithstanding exceptions such as a storm front coming through, the atmosphere at dawn is often as placid as it's going to be for the day.

Thus if birds are going to sing, daybreak provides the optimal acoustic conditions in which to do so. It may be no surprise that they make the most of it. However this doesn't really explain why birds create a dawn chorus. Other animals generally don't, and in theory, birds don't need to. Why a festival of birdsong at dawn?

We can begin by recognising that there is a difference between a bird singing at dawn, and participating in a dawn chorus. For some birds, singing at daybreak fulfils other agendas. Marbled Murrelets for instance, seabirds with the unusual habit of roosting in the redwood forests of America's west coast, call to keep in touch with each other as they fly out to the open ocean to feed each morning. I encountered them while recording in the midst of those lofty trees, hearing their sharp voices way overhead, mingling with, yet quite independent of, the resident Varied Thrushes, warblers and wrens making up the dawn chorus around me. Likewise, parrots, cockatoos and lorikeets will take to the air in great shrieking clouds, wheeling across the glowing dawn sky as they disperse to their feeding grounds. Even some woodland species who roost socially may wake with only a collective bout of throat clearing – in the case of our choughs and babblers, a brief burst of whistling and squawking before shaking their feathers and dropping to the ground to begin their day.

While vocally active at daybreak, none of these birds are participating in the dawn chorus. The birds who do share one lifestyle feature in common; they hold and maintain individual homeranges. Waterbirds, seabirds, aerial feeders and shorebirds are among those who often breed in colonies and share feeding areas, and don't join in a dawn chorus. It is terrestrial birds, particularly the songbirds, who do.

This explains why dawn choruses can be highly seasonal. Homeranges are associated with breeding, which in temperate regions is predominantly undertaken in the warmer months. Elsewhere, such as the tropics where breeding is less seasonal, some expression of a dawn chorus may be heard almost year round, while in arid regions, it may only materialise in response to rainfall.

Patterns in the Dawn

For me, the dawn chorus begins with a magical hour; that time of hushed expectancy in the darkness as dawn approaches. It seems that everything is awaiting the new day. It is a time of handing over. For night animals and birds, it may occasion a last burst of vocal activity, as they give their final calls before heading to roost. This is when one can hear the presence of interesting species such as nightjars or owls, which are more often heard than seen.

Meanwhile, some diurnal birds may already be vocalising, especially if there is a bright moon. Australian Magpies in the outback, fairy-wrens, those White-capped Monarchs of the Solomon Islands, Malabar Whistling Thrushes in India and the appropriately named Spotted Morning-Thrushes of east Africa – these are among the early risers. Various members of the cuckoo family also seem to make a habit of calling early, and their eerie cries can be a feature, particularly of tropical predawns.

All this happens during or even before the time known as astronomical twilight, when the sun is 18°–12° below the horizon and the very first light is paling the sky, so little at first as to be imperceptible.

The next phase of dawn is nautical twilight, from 12° to 6°, when the light grows to the point where one can easily make out one's surroundings. This is when there will come a moment when you sense a quickening as the actual dawn chorus begins.

In Australia's drier regions this can be signalled in an unusual way. White-plumed Honeyeaters live in loose communities, principally across the inland wherever riverine forest or timber-lined watercourses occur. During the day their alarm calls – a loud and insistent piping given by a number of them – will alert other bush birds of any threat. In the predawn, a very similar call from the white-plumes seems to serve a different function. From the silence, they will begin a short burst of piping, one bird stimulating other white-plumes to immediately join in. It only lasts half a minute or so, but seems to initiate the whole dawn sequence, as once given, the white-plumes settle into their actual dawnsongs consisting of sweet, whistled notes. Meanwhile other species seem to take this as their cue and soon commence. I think of the white-plumes as giving the bush wake-up alarm.

From the first stirrings of the dawn chorus, each species will gradually join in, one by one. Once it gets going, it can be difficult to notice when some species commence and others drop out. Imperceptibly, the texture of the whole chorus transforms as this happens, confused further as birds move from one songperch to another or even call on the wing. But somewhere early in the process, the last of the night creatures will have stopped, and the handover to the diurnal fauna be complete.

The duration of a dawn chorus may vary considerably. At midsummer in Sweden, it never goes truly dark and I found birds already singing at 1 a.m. Over many hours, their chorusing gradually swelled as growing light slowly permeated a landscape of mirror lakes and pine forests. By contrast, in temperate regions, outside of the breeding season, there may be little dawn chorus to speak of at all, perhaps just a few sporadic calls here and there. But on average, a breeding season dawn chorus will last something around an hour, more or less.

Eventually, with the same mysterious sense that heralded its beginning, you'll get a feeling of things winding down. This is the final stage of the dawn; civil twilight, concluding with sunrise.[43] By the time the sun appears on the horizon, the dawn chorus will likely have subsided. Some of the later joiners may carry on, constituting a group that segue into the morning birdsong. Or everything may go still – a post-chorus hush.

The dawn chorus may seem a rambling affair, a pleasant cacophony in the half light. With familiarity however, one can recognise structure among the many voices. Around the world, in each environment, there are patterns to who sings, when and how in the transition from night to day. Each species will participate in the dawn chorus by commencing and leaving off singing in its own way, resulting in a somewhat reliable sequencing of activity. Over successive mornings, one can recognise the progression unique to each place.

One explanation suggested for this timing of species is the relative size of irises; that birds with better light vision will wake and begin singing earlier. This seems plausible, and it may be the case among some bird communities. However I've not found it a reliable predictor. In Australia, larger species such as magpies and kookaburras certainly call early, but then so too do our tiny fairy-wrens. Across the continent, various members of the robin family, all small to medium-sized birds, are reliable initiators of dawn proceedings. Similarly, the family of white-eyes, around eighty species found widely from New Zealand to Japan to India and west to sub-Saharan Africa, are among the smallest of birds. Yet in tropical forests, and from Australia's northern mangroves to our bushland at home, I've often found their amiable twitterings to signal that things are getting going. In north America, I was hearing sparrows and small warblers as the first singers.

All this suggests that, while response to light levels may be a factor, to understand the deeper dynamics of the dawn chorus, we need to listen to the singing behaviours of individual species, and appreciate how their participation reflects their sonic strategies.

Showmanship – Northern Temperate Dawnsongs

The dawn chorus has its core species, those that can be relied on to find their favourite spot, be it secreted among dense foliage or a favoured exposed perch, to pour out their song and usher in a new day. In the northern hemisphere, they number some of the superstars of birdsong: Blackbirds, Nightingales, various larks and thrushes, Blackcaps and the European Robins, joined in equally effusive performances by smaller species such as sparrows, buntings, winter wrens and warblers.

These birds are virtuosos, each in their own way. Their songs can be composed of spectacular variety. In the Mojave Desert, surrounded by an arid landscape of yuccas, cacti, sagebrush and iconic Joshua trees, I recorded a Northern Mockingbird giving a standout performance. He was delivering phrase after phrase, each different and delineated by a brief pause. Phrases would consist of a rapid repetition of elements, of either just the one type or several strung together in sequence.

He kept this up almost continually through dawn and into the morning. In one thirty minute sequence, I counted him singing over 450 changes of song type. To be clear, there were recognisable repetitions and so this significantly overstates the extent of his repertoire. Also Mockingbirds are of course famed as mimics so, not being familiar with the local birdsong, much of what he was doing went over my head. The American ornithologist Donald Kroodsma writes of a bird he documented as having around 100 songs in his repertoire, and quotes other naturalists counting upwards of 150 or even 200 in a season. So I'll bow to their more familiar ears.

Yet even my Mockingbird's impressive performance pales beside the Brown Thrasher, which Kroodsma documented singing 4,600 couplets over two hours, and estimates as having over 2000 different songs in its repertoire, many of these being spontaneously improvised.[44]

These extremes of virtuosity are cause for wonder, as I felt listening to Nightingales performing their varied repertoires. On the edge of that abandoned orchard in rural Turkey, ensconced among the brambles and shrubs, I heard a bird singing one phrase after another, each unique and complete. No flubs or fluffs. No hesitancy. He knew exactly what he was doing. Precise and measured, he was a master of his art. As an Australian, I could only listen in amazement to a vocal prowess that has few equals in my homeland.

Given that this impressive singing behaviour is heard from male birds, and most profusely at the beginning of the breeding season, I can appreciate why naturalists in the northern hemisphere would conclude that he is both vigorously defending his patch and competing for a female's attention. That she would choose a mate with the greatest singing skills has been theorised as driving an evolutionary process of more and more complex and bewitching performances. In turn, this has led to a development of the avian brain. In the male it has promoted vocal dexterity, memory and possibly sonic imagination. In the female, a nuanced appreciation of his song that may represent a sense of aesthetics.

That the requirements of holding a homerange and seducing a mate should result in such an evolutionary outcome may seem unarguable. Except for one consideration – as we've previously noted, these elaborate repertoires of dawn-singing birds are heard mostly in the temperate northern hemisphere. Elsewhere, the dawn chorus is different.

Neighbourliness – Antipodean Dawnsongs

On the other side of the world from those Mockingbirds and Nightingales, the antipodean dawn chorus presents a distinct contrast. An example of this difference is heard around the continent as the dawn chorus first begins, with the warbles of Australian Magpies. Unlike northern songbirds, antipodean magpies are not known to sing to seduce. Instead, their carolling duets affirm long-term relationships, not only within their family but with neighbouring groups.

The next contrast comes from various members of Australia's robin family who, across a range of habitats, often initiate the chorus and sustain it through its early phase. The calls of these robins, as we've heard in the *Petroica* genus, are repeated, syllabic phrases. In drier country, it is the Hooded Robin who sings in the dark, often prefacing the dawn chorus well before the White-plumed Honeyeaters have officially announced its commencement. The Hooded Robin's song has a somewhat rough but pleasant texture, a quick *Wii-chaka-chaka-chaka*, which I find highly evocative echoing across open country. 🐦 In forested habitats, the yellow-breasted robins of the genus *Eopsaltria* – from the ancient Greek *eos* (the Goddess of dawn) and *psaltria* (singer) – signal the commencement of the dawn chorus. The Eastern Yellow Robin has a remarkable call for a small bird, an unusually strong *CHAP! CHAP!* These whipcracks resonate through the forest and are repeated well into the full dawn chorus. 🐦 In rainforest habitats, it may be Pale-yellow Robins or Grey-headed Robins that fulfil the same role with similarly simple songs. 🐦 In neighbouring New Guinea, I found a related species, the White-winged Robin, to be the first voice of the cloudforest dawn. 🐦

Listening to these various robins, one usually hears several of them, each calling from the adjacent homeranges on which they reside. They sing strongly, and by doing so, presumably maintain those homeranges and bond with a mate. So they are achieving the same purpose as northern songbirds, yet they don't seem to have got the memo about complex repertoires. Their simple song phrases are replicated one after the other, day after day, year after year, and don't vary audibly among their population. Why? What are they doing that benefits from such a simple repertoire strategy?

Willie Wagtails may provide a clue. A much-loved and somewhat feisty little bird, Willie Wagtails are a familiar resident in suburban parks and gardens. Highly adaptable, they're also found across the continent from rural farmlands through to remote areas of the outback. Their song is a cheerful jumble of notes, often described as *'sweet pretty creature'*. As with the robins, it is a simple phrase repeated frequently and unchangingly, especially during the dawn chorus. However unlike the robins, the Willie Wagtail's song displays noticeable regional dialects.

Dialects are found in many songbirds, presumably because vocal learning allows for the transmission of a localised vernacular. At Mutawintji I was remembering the Willie Wagtail's song as *'you'd better get it, then'*. Dialects could be interpreted as simply an accidental outcome of song learning across geographically distant populations, a kind of 'repertoire drift'. However I think there is more to it than that. Clearly defined dialects enable a new possibility for songbirds: the creation of community. By sharing an identical song, they have the capacity to share a group identity, and thus a belonging. Dialects allow birds to recognise and know their neighbours, acknowledging them through a commonly held song that they broadcast each morning.

That communities of song learning birds may polish and refine a shared repertoire to facilitate living side by side seems reasonable to me. There are benefits to living together, as birds demonstrate by, for instance, forming flocks or breeding in colonies. For an independently breeding pair of birds resident on their own homerange, they could find similar safety and stability by knowing who their neighbours are.

When a new bird arrives, having dispersed to find a homerange of its own, its subtly different dialect would show it to be an outsider. I can't help thinking of a grizzled hillbilly eyeing up a stranger with; "yer not from round here, are ya?" In practice, I suspect that when a bird disperses, it may not lob into a new community and sing immediately. It could arrive and discreetly absorb the local dialect, while listening out for where everyone lives. Once it has identified a vacant homerange, and using open-learning to absorb the local idiom, it may then begin to integrate itself with its neighbours, and soon sound like one of the locals.

Hearing a number of Willie Wagtails singing in the predawn, each giving an identical song phrase in their own local dialect, I sense these birds as living together. By singing, they are acknowledging the presence of others, honouring, negotiating and fine tuning their relationships. Each morning, they reaffirm their place among a known community of neighbours. 🐦

Countersinging – Listening among Birds

A particularly noticeable feature of the dawn chorus is the behaviour of countersinging. Also known as antiphonal calling, it is the alternating of calls between two or more birds of the same species. Rather than singing over each other, neighbours take turns, and one hears them bouncing phrases back and forth. In this way, the silences they create are as important as their songs, allowing space to listen. As much as singing, birds at dawn are being sung to.

Some species are adept at countersinging, while others may sound a little random, occasionally tripping up and singing concurrently. But those pauses to listen are there in many dawnsinging repertoires, as significant to the interaction of birds as the sounds themselves.

My favourite practitioner of countersinging is a master of the artform – the White-fronted Honeyeater. They are medium-sized birds, and relatively common in inland regions wherever flowering plants are found. They're effervescent dawn singers, each individual having a repertoire which sounds like a whole library of mechanical sound effects, such that I think of them as 'the pinball birds'. There are pings, bells, chips, clunks, whistles, tinks and rattles. These are uttered in a quickfire stream of sound bytes, one after another, rarely the same twice, and always with a short gap between: *WEE-chip. ZIOU-tip. ch-ch-ch-ch. Wi-Wi. Chi-dup, Chi-dup. TING! t-t-t-t. tup. zzZIP!* ...

When a second bird joins in, the alternating of their calls can be uncanny. It is a marvellously synchronised performance. With seemingly telepathic empathy, two birds will interlock their songs to create a mesmerising duet. Amusingly, every now and then I may notice one adding a softly exhaled moan, like a tennis player exerting to return serve. 🐦

When a White-fronted Honeyeater sings alone, it has the same repertoire and phrase spacing. However it's rare to hear them solo, as these birds are mostly found in loose groups. In partnership, their singing becomes the interlocking performance it is intended to be. Their repertoire assumes its relevance, uniting neighbouring birds in the execution of a recital which may continue without pause for up to an hour.

White-fronted Honeyeaters are far from the only species adept at countersinging. The White-eared Honeyeaters that live around our home – our 'kWHI-choo' birds – are also pretty good at it. Their dawn chorus is far simpler than that of the White-fronted Honeyeaters, a single repeated phrase sufficing: *kWHI-choo, kWHI-choo*. These are given every few seconds, while a neighbouring bird replies.

The more you listen closely to a dawn chorus, identifying individual conversations among the many, the more you will pick up these alternating patterns and their associated silences. In doing this, your ear is attuning to that of the singing bird as it listens out for its kin, acknowledging and working in with them, rather than performing to outdo them, or as if they didn't exist. Countersinging is no accident or coincidence of timing. It is intentional and refined, an essential protocol upon which avian dawn negotiations are conducted.

Counter-calling – Listening among Frogs

I'll take a brief detour here, and discuss another group of animals well known for their sense of timing; the Anurans – the frogs and toads.

Male frogs call to advertise for a mate, although in some species female frogs may also call in response. Their voices are amplified by vocal sacs or membranes of skin under the throat, which balloon in quite spectacular fashion. The amplification that skin membranes allow is both integral to the vocalising process, and remarkably effective.

The volume of a frog's call has been suggested to be an indicator of reproductive fitness. That may be a part of it, but there are other purposes of amplitude. Think of it from the frog's perspective. They are calling from the ground, the water surface or low vegetation at a pinch. They need a good bit of oomph just to get the sound out. This is especially noticeable in some small tropical frogs that create ear-piercingly loud 'peep's to carry through the acoustically absorbent rainforest ground layer.

Those species that live in water bodies have additional need of loudness, because mating with nearby frogs is not ideal. A healthy population requires genetic diversity, and that necessitates population exchange from habitat that may be some distance away. Whilst their little amphibian brains will not be aware of the genetics involved, their behaviour has been naturally selected by ancestral frogs that have attracted their mates from far afield. Thus they call loudly, not to outdo rivals, but so that their community can be heard across the landscape. Counter-calling facilitates this, as each individual animal contributes to maintaining a continual thrum of sound in the air.

The result is that frog choruses may be audible over great distances. One can often map water in the landscape at night by simply hearing where frogs are calling from. When you do, you're listening like a frog. Migrating frogs are similarly listening for where the water is, following sonic maps created by others of their kind to navigate hostile terrain in search of new homes and mates.

I find it fascinating that frogs have optimised their calling patterns to create the most sound, depending on the number of vocalising animals. When only a small number of frogs vocalise together, counter-calling may be executed with considerable precision. At the front of our home is a pond which is home to Brown Froglets. Recently, after rain, a pair could be heard calling sporadically throughout the night. One would begin with a *crick, crick, crick* ... and quickly the second would join, filling the gaps, the two establishing a pattern of perfect reciprocation. They called like this for a minute or so, but as soon as either stopped, so would the other. This process was repeated throughout the night.

Brown Froglets are tiny, half the size of your little fingernail, which amplifies my amazement at the sophisticated interplay they display with each other. 🐦

When a larger community of frogs chorus, this precision of counter-calling breaks up into far more ad-hoc patterns. Whereas a duo or trio of frogs may be heard responding to each other in some way, once a few more join in, it all goes to pot. The community begin calling randomly, in a manner that reminds me of the haphazard fall of raindrops. And who knows, perhaps they really are calling up the rain gods? This random temporal patterning is termed stochastic, and is found throughout nature in various forms, frogs being very audible exponents of it. 🐦

No matter the number of vocalising frogs, a continuity of sound is created. Once a potential mate is attracted by this throb of sound, another feature of frog chorusing comes into play. Neighbouring animals will be calling antiphonally or somewhat randomly. Either way, as with birds, they're not signalling over the top of each other. Hence a female frog can more easily select a potential mate from among the general chorus.

Another benefit of anuran antiphonal calling may be that it serves a defensive purpose. By keeping an undifferentiated wall of sound in the air, a predator will find it confusing to pick one voice from among the many. By the time it is close enough to target an individual from the many, the vibration of its approach will likely have silenced the whole frog community.

Thus the chorusing of frogs represents many things; mating, yes, but also navigation, protection, genetic resilience and perhaps a degree of celebration at the arrival of rain.

Belonging – Honeyeater Dialects

Returning to birds – honeyeaters had yet another thing to tell me about how the dawn chorus functions, and they'd been telling me from the very beginning. What I loved about those Spiny-cheeked Honeyeaters at Mutawintji was the slowly descending tonality of their piping songs. As I've recounted, I found them exquisitely beautiful. 🐦

Years later I visited Gundabooka National Park, about 200 kilometres east of Mutawintji. An extensive area of woodlands centred on a dramatic syncline of outcropping hills, it also has a healthy population of spinys, and I was looking forward to hearing them sing at dawn. However I was to be both disappointed and puzzled. The Gundabooka spinys didn't have that same tonal quality at all. In place of piping notes, they gave a rapid series of chips on a steady pitch. They didn't possess that sublime vocal texture of the Mutawintji birds. I was hearing a different dialect. 🐦

On another trip, Sarah and I spent several days exploring an ephemeral wetland we'd chanced upon amidst otherwise barren country, 200 kilometres north of Gundabooka. The area was flooded by exceptional rainfall and numerous species had flocked in to take advantage of the sudden wetlands. There were flotillas of Pink-eared Ducks out on open water, Budgerigars breeding everywhere, dotterels patrolling the water's edge on dainty legs, and Bourke's Parrots fluttering in under cover of dusk to drink just before nightfall. It was, it's safe to say, a happening place.

Rising early, I found Willie Wagtails were already beginning the dawn chorus. As the chorus grew, I noticed a voice I couldn't identify at all. I'd never heard this song before – a curious pulsing of buzzy notes. As it repeated every now and then, I was going through a checklist of possibles in my head and drawing a blank. It was only later in the day, watching dozens of spinys raiding the abundance of flowering eremophilas, wattles and eucalypts, that I realised that once again, the species had tricked me. The dialect of spiny dawn song in this place was so different that I hadn't recognised it at all. 🐦

Nearly twenty years after those Mutawintji spinys had first sung to me, I returned to climb that same ridgetop before dawn and hear them again. They would be the next generation, or even two, from the birds I'd first recorded, and I was curious to hear if their song had changed. As a pink flush appeared in the sky, I felt that I was back home among friends. Here was that piping song just as I'd heard it all those years before. They were still singing the landscape with the same dialect – the song of Mutawintji.

Some time later, I took a detour to check out that ephemeral wetland again. I found the place completely dried and desolate. Where there had been open water was now a cracked claypan, and my footfalls kicked up only dust. I found no birds, they'd all moved on elsewhere. In place of abundance was abandon.

In all these dawn encounters, spinys were telling me a story. Sometimes they are resident in habitats that sustain permanent populations, Mutawintji and Gundabooka being two such oases in otherwise semi-arid country. There, spinys have their own local dialects, and pass them on, to be heard from one generation to the next. They are the signature dawnsongs of that population. They signify belonging to that place and community.

The spinys of the ephemeral wetland however were not a resident population. They may have arrived opportunistically from further afield in the region, or been a perennially nomadic band of birds that move vast distances in response to rainfall and flowering. Who knows from how far away they had travelled? Yet they nevertheless had a dawnsong that united them, like a band of marauding pirates gathering under the skull and crossbones. Recalling them meticulously raiding the flowering eremophilas, that analogy somehow seems appropriate.

Spinys have told me another significance of the dawn chorus. Only in their dawnsongs can one hear these bird's expressions of identity; nomadic or resident, and to which community they belong. You can't determine it from a spiny's daytime repertoire, which is completely different. A whiney, whingey concoction of weedling squawks and gurgles, it is as removed from sublime as you could imagine. This diurnal repertoire is also fairly uniform and characteristic of the species, heard wherever spinys are found. It seems to facilitate daytime interactions, rather than speak of any regional or group affiliation.

A special dawn repertoire with pronounced dialects is not unique to spinys. Many of Australia's dry country honeyeaters possess them similarly. White-plumed Honeyeaters are an example. After waking the bush with their piping alarms, a local group will commence singing with dawnsongs that are sweet and distinctive, often a simple pair of down-sweeping whistles, *Sieup, Sieup.* 🐦 But as one travels around the country, you can recognise local variations wherever you go. 🐦🐦🐦🐦🐦 As with the spinys, the daytime calls of white-plumes are completely different, and relatively consistent across populations. 🐦

To spinys and white-plumes can be added a whole fauna of inland honeyeater species with unique dawn repertoires, among them Singing, Grey-headed, Yellow-plumed and Grey-fronted Honeyeaters, 🐦🐦 plus our 'kWHI-choo's, the White-eared Honeyeaters. These species are resident where conditions are suitable, or locally nomadic. But when droughts deepen, or rainfall creates temporary opportunities, they will move sometimes great distances in response to localised flowerings. When doing so, they join others such as Pied and Black Honeyeaters, whose perennially nomadic movements across the inland are far more unpredictable and mysterious.

The implications of this may reach beyond the vocal behaviour of the birds themselves. For many of Australia's desert plants, honeyeaters are a principal pollinator. Forming associations with specific plants, honeyeaters fulfil a vital function in ecosystems across the continent. In their movements they carry pollen, transferring it from one region to another. If honeyeater dawnsongs allow us to discern the communities and roving bands to which they belong, then by acoustic monitoring, we can potentially track their movements across the landscape. This may give us a measure of the dynamics of arid ecosystems across vast areas of the continent. In addition to telling us how birds negotiate their lives together, the dawn chorus may hold a key to understanding the productivity and stability of our desert land systems.

Sequencing – New Zealand's Native Nectarivores

The dawn singing of honeyeaters had an additional surprise for me, one which amplifies the significance of all I've understood of their behaviour so far.

New Zealand is the only country where Sarah and I have found people making a special effort to rise early just to listen to their native dawn chorus. This is possibly because their birdlife has been so devastated by the ill-informed introduction of non-native wildlife. Since European colonisation, a plague of cats, stoats, possums and rats have exterminated or drastically reduced the populations of many native birds on the mainland. Dismayed at the loss of their distinctive birdlife, New Zealanders have formulated an ambitious program to eradicate these pests, restore their lands and reintroduce their native species over coming decades.

In the meantime, the best place to hear the mournful voice of a Kokako, the perky calls of Whiteheads or the strident songs of Saddlebacks is on an offshore island. Some have been specifically set up as bird refuges, and one of the longest established of these is the small island of Tiritiri Matangi.

The day Sarah and I arrived, we reconnoitred and I found a patch of forest that seemed a good location for recording. As it was some way distant from our accommodation, I wanted to make sure I had adequate time to walk there. Rising early the following morning, I shouldered my gear, mildly surprised to find others in our shared bunkhouse also stirring. Stepping outside, I realised why. The Tuis were already vocal, their curious 'yoik's and twitters drifting from copses of trees on the night air.

With the lights of Auckland glowing across the water, I set out across the island in the dark. As I walked, I reflected that in Australia, the recording aesthetic I developed early on often came from hearing the countersinging patterns from a single honeyeater species. I'd often found the cohesion of the dawn chorus hinged on the presence of one, or occasionally two (usually a larger and smaller) species. Even when multiple types of honeyeater were in the district, it seemed the dawn chorus in any particular locale was usually orchestrated by only one.

This was a puzzle that deepened when we moved into our bushland home twenty years ago. At first, Yellow-tufted Honeyeaters were our resident honeyeater. They were so ubiquitous and numerous that we got to christening them 'the bush mafia'. Dozens would flock to our birdbath in the afternoons, and the dawn chorus was alive with their chiming whistles. Then, about seven years ago, we noticed there seemed to be less of them, and eventually there'd be months during which we didn't see any at all. They remained in the district, just not around our place. Instead, Yellow-faced Honeyeaters first moved in, then Fuscous Honeyeaters. Both moved on in turn, and in recent years, White-eared Honeyeaters, previously only an occasional winter visitor, have been resident in the garden year round and wake us every morning.

At first, I thought that local wanderings of honeyeaters were responsible for the changing dawn chorus. Then I began wondering if it weren't the other way round – that the dawn chorus was moving the honeyeaters. Or to be more precise, communities of each honeyeater species were roosting apart so as to sing separately the following morning. By doing so, their negotiations of the dawn could be more efficiently achieved, as each species was singing in their own place in the landscape.

Now those experiences of hearing single honeyeater species at daybreak all around the country began to make sense. Honeyeaters were geographically organising themselves to facilitate their dawn chorus.

As I walked, I was thinking about New Zealand's three native nectarivores: the larger Tuis, the more medium-sized Bellbirds, and the slightly smaller Hihi, or Stitchbirds. They were each present on the island, thanks to reintroduction programs by tireless and passionate volunteers. Feeding stations had been set out to help them establish, which has resulted in a population of each that is larger than the natural carrying capacity of the island. And Tiritiri is small, only two kilometres long and not even one kilometre wide, so the birds were packed in. In light of the behaviour of Australia's honeyeaters, I wondered how, with so little space to spread themselves out, New Zealand's three native nectarivores would handle their dawn chorus.

As I set up my microphones, the air was calm and I was hopeful of a good morning's recording. We'd learned that we were ridiculously fortunate in the timing of our visit. It was spring, but apparently the weather over the preceding six weeks had been foul. Blowing a gale continually, the birds had been hunkered down and unable to get on with much. This was the first day when conditions were to be still and perfect. Like puppies in a park, they were about to be let off their leash.

The Tuis were already off and running. From the few birds lazily calling as I set out, there were now dozens of them to be heard in the dark. It seemed the whole island had become adrift with Tui song. Switching on the recorder, I stepped back to listen.

I had lucked the right spot. From the branches above me, a Tui began serenading, allowing me to hear the complexity of its song close up. How to describe the vocalisations of Tuis? I could hint at it by saying that they sing with one of the most extensive frequency ranges of any songbird. From low notes below 400 Hz, their voices extend through the frequencies to almost inaudibly high. Like mythical creatures, they sing with a voice both deep and thin, strong and sensitive, harsh and sweet, producing a tumble of sounds that seems infinitely varied. Articulate and playful, a single song bout of 10–15 seconds may be composed of thin whistles, a grating outburst, soft wheezes, clicks, whinnies, explosive utterances, silvery notes and all ending on a final, soft chuckle. Interspersing these bouts they'd add a loud *Yow, YOWK!*, refrains of which I could pick up from faraway birds. 🐦 I'd really underestimated these Tuis. I guessed they'd already been vocalising for well over an hour, and in that time, the intensity of their singing had grown to the point where it seemed the whole island was ringing with Tui song.

Then, as first light began creeping into the sky, I became aware of a new voice emerging. Almost subliminally, delicate whistles had begun to form tinkling, arpeggiated patterns, cycling over and over, gaining in presence. The Bellbirds had entered the dawn chorus. ♪

With this, something remarkable happened – the Tuis began quietening. Over the following minutes, the Bellbirds became more and more audible, while the Tuis fell progressively less vocal. In a little over five minutes, the transition was complete and the Bellbirds now had the stage. For the next half an hour, their chiming voices came from every corner of the forest, creating an effect that was enchanting. This must have been the sound that Joseph Banks, botanist aboard Cook's *Endeavour*, heard as their ship was moored offshore, and of which he wrote in his diary of January 17, 1770: "their voices were certainly the most melodious wild musick I have ever heard, almost imitating small bells, but with the most tuneable silver sound imaginable." ♪

As the Bellbirds continued, other native species began awaking; *zii,zii,zii,zii,zii,zii*, that was Popokatea, the Whiteheads. A pleasant tinkle of notes would be Piwakawaka, the New Zealand Fantail, sounding similar to the species we had at home. And Toutouwai, the New Zealand Robin, a member of the *Petroica*, but with a surprisingly more extensive repertoire than the Australian members of its genus.

After some time, I noticed the chorus of Bellbirds beginning to diminish, and a new voice appearing. It was a high-pitched, sharp, almost spitting call. This was Hihi, the Stitchbird, both Maori and western names deriving from its voice, and the third of the native nectarivores. ♪

As the Bellbirds ebbed away and the Stitchbirds became ever more voluble, there was another graceful transition from one species to the next. This passing of the baton also took a few minutes, by which time the air was filled with the Stitchbird's bright, incisive *Tst, Tst, Tst* … They remained calling for the next half an hour, quietening as the sun rose, when other species such as the Saddlebacks and the sad-voiced Kokakos eventually awoke. ♪

Over the following mornings, I heard the same pattern repeated. The Tuis would begin singing in the early hours, gradually intensifying through astronomical twilight. With nautical twilight, and as the first tones of the Bellbirds began swelling, the Tuis would quieten. The Bellbirds would chime for the next half an hour, to be followed around civil twilight by another short transition as the Stitchbirds took their turn to usher in sunrise. The entire dawn chorus lasted over two hours, with clear phased transitions between one species and another.

To hear this sequencing was fascinating, a demonstration of how these native nectarivorous species could organise themselves in the dawn chorus. It implies some degree of interspecies familiarity, but with a twist – Stitchbirds are not actually related to honeyeaters. They have recently been allocated their own family, likely descended from an earlier radiation of passerines to New Zealand and sharing origins with Kokakos and Saddlebacks. What they do have in common with Tuis and Bellbirds is their enthusiastic raiding of the island's nectar stations, and perhaps this dietary niche is enough to have shaped their place in the dawn sequence.

However their taxonomy suggests there may be more to it. Stitchbirds eschew higher-quality food sources when the more pugnacious honeyeaters are present. Could an un-honeyeater-like timidity be responsible for them only singing once the honeyeaters have concluded? Or, rather than last, as I initially assumed, could the Stitchbirds actually represent the vestige of a far more ancient sequence with their cousins, the Kokakos and Saddlebacks, who follow them to become vocal after sunrise? Even more intriguingly, could Stitchbirds be bridging the two – maintaining ancestral patterns while also fitting in with the more recently arrived honeyeaters?

Whatever is shaping the dawn sequencing I heard on Tiritiri Matangi – whether a local response to the density of nectarivores or a long-established behaviour among native species – I can only surmise. Sadly I was not to hear it repeated elsewhere, as the numbers of them on the mainland, particularly the Stitchbirds, were so low. Perhaps if the restoration efforts of New Zealanders are successful, and these birds are returned to the native haunts, we may one day find out. 🐦

Throughout our journey around the mainland islands of New Zealand, I recorded other dawn choruses with Tuis prominent, but with Bellbirds and Stitchbirds absent. Without that interaction of fellow nectarivores, the Tuis sang alone, evocatively but without constraint, continuing until well after sunrise. The dawn chorus had lost its structure, an integrity that only a full suite of native species could provide. 🐦

Song Mirroring – Grey Shrike-thrushes

Over recent years, a pair of Grey Shrike-thrushes have nested under our carport. To be honest, it wasn't their initial choice. They first set up home inside the adjacent shed that houses our solar power system. They sneaked in one day when I inadvertently left the door open, and before I realised, had built their nest, a loosely woven cup of bark strips. Having established themselves, we didn't have the heart to evict them. The door was finally closed only after their chicks had fledged and were piping demandingly from the roof rack of our car.

The following year, having not learned my lesson, they returned through a casually unattended doorway. When I noticed them flitting in and out, it was to tend three freshly laid eggs. Enough is enough, I said to them, and constructing a chicken wire sling and attaching it to a beam under the carport, gently moved the entire nest, eggs and all, outside. We watched anxiously as they returned within minutes, inspected the new arrangements, and settled straight back down to incubate. They've returned there every year since.

Grey Shrike-thrushes are wonderful singers. Beloved by many Australians, their melodic songs are given in a voice that is rich and tonal. Characteristically beginning with a few soft lead-in notes, a song phrase will consist of a strong ripple of tuneful notes flicked into the air, often concluding with a rising note or occasionally a brief rattle. An individual will have possibly several dozen song phrases in its repertoire, each unique and recognisable.

During daytime singing, a bird will usually repeat one song phrase at leisurely intervals for a few minutes before taking up another. But every now and then, I'll hear the bush come alive with shrike-thrush song in a far more intense and extrovert display as a bird utters a variety of ringing phrases in quick succession. Catching sight of them, I'll see two or more individuals, tails cocked and wings aflutter, hopping from branch to branch while one sings in animated voice. Whether this is a mating display, bonding between a pair, or even a negotiation between neighbours, I've yet to conclude. It may even fulfil various purposes depending on the circumstances.

Grey Shrike-thrushes are found widely across the country, and everywhere one hears their songs, while their voice is immediately recognisable, the specific phrases they have are different. Being a song-learning species, it could be assumed these are expressions of regional dialects, however this doesn't seem to be so. Instead, each local community of birds can be heard to share a distinct repertoire of song phrases in common. Even short distances or changes in geography, such as between ridge country and neighbouring river flats, or one valley and the next, can result in a noticeably different collection of songs.

Independent of these local shrike-thrush repertoires, there appears to be one call that is universal to the species, an explosive *PEE-OO!* This has been suggested as a contact call, and I've certainly heard them giving it when calling back and forth to each other. However I believe it may have multiple significances. On occasion, we hear a single call echoed back from the bush in response to a slammed door or other sharp disturbance. They'll also give repeated, urgent *PEE-OO!*s in distress at a threat such as a kookaburra. And sometimes it seems to be uttered for no discernible reason at all. I've come to imagine this call as a shrike-thrush's version of an expletive. If so, while shrike-thrush song is somewhat seasonal, they seem to cuss all year round.

The shrike-thrushes nesting under our carport clearly believe they own the title deeds to our property. Not only have they set up house, but their homerange encompasses a good area of the surrounding bushland. We think of them as 'our' thrushes. The male is a rich singer at dawn, waking us over the years with his strong and lyrical voice. However as I became more curious about the dawn chorus, I began wondering what he was up to exactly. I felt intuitively that he was doing something interesting, I just couldn't put my finger on what it was.

So, during the course of a week one spring, I put my microphones out in the garden each morning to eavesdrop on our resident shrike-thrush. What I heard him doing has revealed to me another unexpected layer of interaction in the dawn chorus. Here I shall describe just one of those mornings.

The dawn chorus started in its usual manner, with a yellow robin, magpies and a group of kookaburras, all calling from some distance off. The fairy-wrens seemed to be waking a little later than usual, but eventually began in fine voice, along with a fantail and Rufous Whistler. By this point, the eastern sky was getting light, and the dawn chorus had been gaining intensity for a quarter of an hour.

It was at this point that 'our' shrike-thrush began singing by giving a single song phrase from his extensive repertoire. He had to wait another full minute before a reply came from his neighbour, a shrike-thrush on an adjoining homerange perhaps 300 metres distant. When that reply came, it was with exactly the same phrase. Intonation perfect, it was identical.

Our shrike-thrush immediately responded, repeating the phrase. Once again, another wait. His neighbour was a bit slow getting going, only calling every so often. But after a few minutes of adjusting their timing, they both settled, each singing every ten seconds or so, and each mirroring the same song. 🐦

This continued for ten minutes or so, and then – a change. Our shrike-thrush sung an entirely different song phrase. There was a pause from the neighbouring bird. When it sung again, it was to echo this new song. The two shrike-thrushes soon fell back into countersinging, but now exclusively using this second melody. Meanwhile another bird had begun from further away. It too was using the exact same phrase. For the next ten minutes, this little group of birds countersung beautifully, passing that one song between them. ♪

During this time, other shrike-thrushes could be heard singing across the expanse of the landscape, some barely discernible from far away, but all utilising this one song. Presently, our bird offered yet another new phrase. However a magpie happened to sing at the same moment, obscuring his voice. He reiterated his fresh offering, but again got tangled with the Magpie, while his neighbour stuck to the earlier song. Our bird then fell quiet, as though listening. But another change was afoot from further away. A distant bird had introduced its own new phrase, and now the neighbour was picking up on that one. Our bird rejoined with the latest song, and in no time at all, the shrike-thrush neighbourhood were synchronised around this novel motif, passing it back and forth among them. ♪

By now the dawn chorus was abating a little, and with numerous shrike-thrushes singing, the next change was a little more confused. A fourth song was faintly heard, and our shrike-thrush took it up, matching it perfectly. However the neighbour and another didn't, and continued as before. Seemingly confused, our bird paused, momentarily returning to the previous song. However the others then caught on, and shortly the new song was taken up by everyone. Our shrike-thrush adjusted, and once again, everyone fell back into sync. ♪

During the next cycle, our shrike-thrush suggested new song phrases twice, but neither were taken up, and on each occasion he returned to the collective one. By this time the dawn chorus really was waning, and with it the number of shrike-thrushes singing. Gaps between songs were lengthening, and a handful of minutes later, everyone fell quiet. ♪

Sunrise nearly upon them, and their dawn negotiations complete, the shrike-thrushes in our district got on with their day.

I soon noticed our bird giving his more usual diurnal repertoire of varied phrases as he began moving off to forage for breakfast. There was even a *PEE-OO!* to be heard.

In listening back to my recordings of our local shrike-thrushes, one thought kept going through my mind. This was the bird nesting under our carport. It is not a co-operative breeder. It doesn't live in communities like our honeyeaters. Shrike-thrushes breed as a mated pair, tending their nest as a couple and maintaining their homerange in the adjacent bushland. Their singing should be a textbook example of a bird defending its territory. That birds should sing rather than fight already undermines this picture. The aggressive implication of a term such as 'defence' doesn't feel appropriate for a practice in which birds all over the world use singing to negate the need for physical confrontation. Communication, negotiation and dialogue – not dominance.

But our shrike-thrush's singing behaviour at dawn goes beyond even this necessity. Unlike northern songbirds, he is not simply broadcasting to his neighbours or advertising to his mate, he is participating in the creation of a collective performance with other members of his species beyond his immediate boundaries. In the cool and still atmosphere of dawn, I was hearing him respond to birds from a kilometre or more distant.

Shrike-thrushes also transcend my understanding of counter-singing. More than sharing a simple song, such as robins and Willie Wagtails, or alternating it with neighbours as honeyeaters do, they are adding an additional layer of complexity. From their extended repertoire, they are selecting individual song phrases, and utilising them as the basis for their interactions. If they were only selecting one phrase, I'd think of it as a more specific version of conventional countersinging. But when that phrase is changeable and negotiable, it brings an extra dimension into play. Birds must listen more actively, attentive to changes coming from their wider community. They have to decide how and when to respond. They may also initiate. If a bird's initiative is not responded to, it will be abandoned in favour of following the collective. This cannot be a formulaic practice. It has to involve some degree of decision making on the part of the birds. They are weighing up priorities such as the novelty of a new song versus the desire to fit in with their neighbours. Once again, this suggests we are hearing a perception of self and other.

I hear the shrike-thrush dawn chorus as a complex interplay, a game of relationships. It could be tempting to say it is a game of dominance, but I don't believe it is. Transitions between song phrases ripple through a community organically. Birds respond to them promptly and always. They also contribute by initiating, sometimes successfully and at other times being bypassed. No, this is not a hierarchical interaction, but one among peers. Each bird's voice is heard, and has a role to perform in the shrike-thrush chorus.

In the process, our shrike-thrush was not only communicating with its neighbours 'over the back fence', but with the wider district, the whole suburb if you will. They were creating their own collective performance, which spanned the acoustic space, possibly beyond the limits of immediate listening distance. I suspect that greater distance would generate sub-communities of singing. If one happened to listen on a boundary, one would hear a discontinuity between the interactions of one population and those adjacent, like standing between two cricket pitches. From my recordings, I suspect those boundaries shift significantly morning to morning, rather than being indelibly 'drawn on the landscape'. If so, this adds an additional layer of plasticity and challenge, with individual birds not being tied to unchanging communities. The play of the game just gets more and more complex.

That shrike-thrushes should be performing this song mirroring ritual suggests to me a form of culture. Not only are they learning their local repertoire of song phrases, but they are then employing these to engage with their neighbours. There is likely to be an element of pleasure in this for the birds, their brains responding to neurochemical responses as they parse the state of play and contribute to it. If so, then we can expect that, as well as participating in a process vital to their lives, they're also enjoying themselves.

The sophistication of their behaviour prompts further questions. Shrike-thrushes are archetypal songbirds, so they're unlikely to be a lone practitioner of such refinement. Rather than unique, their song mirroring will be but one expression among many ways that songbirds interact in the dawn chorus. If shrike-thrushes are engaging in an interplay of such nuance, what other wonders are yet to be discovered in the dawnsinging of other species?

I now listen more attentively to our local shrike-thrushes during the days. The female sings as lyrically as the male, and I frequently hear them trading song back and forth. Sometimes they synchronise and mirror each other perfectly, and at others they each stick to their own song phrases. Occasionally a neighbour will even join in mirroring. What are they playing at in these exchanges? I suspect they will keep me listening and trying to work them out for some time to come.

We awoke one morning recently to the sound of urgent *PEE-OO!* calls from outside. Hastening out, I was just in time to see a Brown Goshawk fly off, and find our two shrike-thrushes flitting from tree to tree in an obviously agitated state. Their nest, despite being hidden under the carport, had been predated, with all three newly hatched chicks taken.

We anticipated this would curtail their breeding efforts for the season, but a few weeks later, the female was back on the nest. Feeling remiss that I hadn't responded quickly enough earlier, I had a word with her; "you tell me if that old goshawk comes back, and I'll sort it for you". I know; leave nature alone, don't interfere, the goshawk is a magnificent animal and needs to live too. But we'd become fond of our shrike-thrushes.

A week later I came out one morning to find her off the nest. Thinking she was off foraging, I was about to walk on when I noticed drifts of soft grey feathers and breast down littering the ground, snagged on twigs and fluttering gently in the breeze. The goshawk had been back, and this time she'd been taken.

I saw the male later that morning, perched high in a nearby tree with a small cicada in his bill. Normally he would have flown into the nest directly with it, but now he remained, calling occasionally. Was he calling for his lost mate? Shortly after, I found him perched on the side of the nest, gazing down silently. Every now and then he'd look up, staring out. I'd never seen him so still and passive. I couldn't help feeling sorrow for him, and wondered whether he was grieving. Do birds grieve? I can't see why they wouldn't. When changing over at nest duties, they'd often whisper quiet, sweet whistles to each other in a repertoire very different to their 'out in the world' calls. If their lives were brought together in emotion and shared endeavour, why wouldn't he feel loss? I watched him for over twenty minutes as he remained there, looking very alone.

I also felt sad at the loss of his beautiful song, as I anticipated this would silence him for the season. So I was surprised when, the following morning, he was singing as strong as ever. Was he calling out for his mate, summoning her one last time? Was he advertising for a new mate? Was he comforting his solitude by connecting with his community? Whatever, the season was moving on and soon he was less vocal. I've since seen him perching near the nest, suggesting that when he finds a new partner, he'll be back next season, to reconnect with his neighbours and fill our bushland with song again.

Messenger Birds and Spark Birds

My friend Jennifer Ackerman, the American science writer and author of *The Genius of Birds* and *The Bird Way*,[45] was the one who introduced me to the idea of a 'spark' bird; that species which first fires one's interest and leads to a lifetime of fascination in all things avian. "What was yours?" she asked me shortly after we first met. "*Passer domesticus*," I replied, the House Sparrow. Uninspiring, but true. An introduced species; common, dull brown, and ubiquitous to the point of sometimes being regarded a feral nuisance.

I must have been around twelve, and already had a bookshelf populated with titles on dinosaurs, whales, beetles, butterflies … so I was ready for birds. One day, I was looking out the kitchen window, idly watching the sparrows that inhabited our garden in suburban Sydney. My mother, possibly seizing on anything she could think of to occupy a bored youngster during long school holidays, suggested I try drawing them. I still have that first sketch of a female House Sparrow, and even in that tentative and awkward drawing, I can see my attention to details of plumage. As the holidays wore on, I sketched more of the birds that appeared in our garden, getting to know each by doing so.

A year or so after this, I returned from a bushwalking excursion having seen a bird I'd not encountered before. It was medium-sized and quite distinctive; olive green above, lighter below, with a lemon-yellow head, a striking black 'Zorro' mask through the eye and distinctive ear tufts of yellow feathers. As soon as I arrived home I sketched it from memory, but despite a search through my bird books, could not identify it.

The Australian Museum is one of the southern hemisphere's great natural history institutions, and my mother took me to visit regularly. On our next outing, she suggested I bring along my drawing, to see whether someone could identify my mystery bird. We asked at the reception counter, and a telephone call was made. A quiet and somewhat old-worldly gentleman emerged from the bowels of the building, and introduced himself as John Disney, the curator of birds. He viewed my drawing, and beckoned us to follow him. Down long corridors, up flights of stairs, past book-lined offices, deep into the back section of the museum we trailed him, arriving at a large room packed with floor to ceiling metal cabinets. Threading our way between them, John eventually stopped and opened a set of double doors, revealing row upon row of drawers. He slid one out, and there, lying on their backs in rows, were dozens and dozens of taxidermied bird skins, each with a handwritten label tied to its feet by string. The whiff of mothballs was overpowering. Picking one up, he turned it over in his hand to reveal distinctive yellow ear tufts. "Is this what you saw?" So, yes, a Yellow-tufted Honeyeater, *Meliphaga melanops* (as it was known then). The skins had been collected from localities across Australia's southeast, some specimens dating back to the 1800s. In his hand, John held the actual 'type specimen' from which the species was first described.

I suppose my enthusiasm was evident, because once the extent of my drawing interest was revealed, I was invited to come back anytime and use the specimen collection as reference material. And so, every school holiday break over the following years, I'd set off for the museum. I preferred to draw a bird from life, and never embarked on a species I hadn't seen in the wild. But having the specimens before me allowed far more accuracy with plumage details. With John's ongoing encouragement, I became a reasonably accomplished ornithological artist, contributing drawings to the scientific papers and museum publications he and his colleagues were preparing.

However this youthful interest was not to last. Having sub-mitted a folio of my bird drawings for the art component of my final school examinations, I was disappointed to receive a poor mark. Art was the subject I felt most enthused by, yet my abilities had been judged insufficient. I came to the saddening realisation that my work was simply not seen as art. It was illustration. Art was something else. It seemed nature, especially rendered accurately, simply didn't qualify.

To be honest, I'm glad I didn't pursue scientific illustration further, as my life took different paths thereafter. Over the ensuing years, I travelled widely overseas. Eventually, arriving home in Australia, the first thing I did was go hiking for a few days. I wanted to reacquaint myself with the Aussie bush. I found a quiet spot in the heathlands north of Sydney, where a small waterfall trickled over a ledge. I stayed there for a few days, sleeping under familiar southern stars, and wondering what I was going to do with my life.

At this point, I had no obvious career direction, and felt at a complete loss. In the sunshine each morning, a big goanna who evidently lived under a rock ledge opposite, would come out and bask. We had some good discussions, that goanna and I. By the time I packed to walk out, we'd decided that whatever I did, it wasn't going to be an 'off the shelf' job. By following my interests and abilities, I'd try and find my own vocation.

Fast forward five years. Sarah and I had met, and with borrowed microphones and a somewhat hazy plan, set out for Mutawintji. There, my knowledge of Australia's birds resumed its relevance, helping me to identify the species I was recording, including my mystery predawn singer.

Looking back, I can recognise birds have appeared at key moments to guide me. One of the world's most ubiquitous birds initiated my childhood interest, a Yellow-tufted Honeyeater deepened it, and those Spiny-cheeked Honeyeaters eventually invited me to listen.

Many years later, back in central Australia, I was assisting an Aboriginal community with an acoustic monitoring project. The indigenous rangers, Wayne, Russell and Kleon, spoke about nyi-

nyis and other species, referring to them all by their local language names. They told me of the relationships Aboriginal people have with each. Jinti jinti, the Willie Wagtail, with its cheeky, confiding manner and the ratcheting call after which it's named, is a messenger bird. "It will tell you things that one – important things." Again, I recalled Harold; if nature trusts you, it'll talk to you.

We all have a relationship with nature, one that, perhaps without noticing it, has been a gentle influence at important moments. Perhaps nature has spoken to you in this way. It has to me – sparrows were my spark, honeyeaters my messengers, and a goanna my career counsellor.

Chapter 11

The Listening Peoples

The Early Australians

It was a swelteringly hot morning, the kind you wonder how long you can stand out in the open before beginning to wilt. Sarah and I were in the far northwest of Australia, in early summer, with coastal humidity adding to the day's growing heat. However, we weren't going to miss visiting this place.

We'd gathered at a carpark with a handful of other visitors to view the world's most extensive collection of early human rock art. Murujuga, or Burrup Peninsula, is an open air site where Aboriginal peoples have carved petroglyphs onto rock surfaces over tens of thousands of years. Some of the art has been dated back 45,000 years.

In Europe at this time, *Homo sapiens* were still interacting with the last Neanderthals. In Australia, *H. sapiens* had arrived and were creating culture in a landscape markedly different to the one we now assembled in. For a start, during glacial periods the sea level would have been over one hundred metres lower, with the place we now stood being inland some 120 kilometres and surrounded by a broad, coastal plain. It would also have been lush and a lot cooler. I could only wish.

Our local Aboriginal guide gathered us together and led the short walk over to a low hill composed of chaotically arranged rocks. It looked like a giant had raked the landscape into ridges of rich, chocolate-brown boulders, their iron content oxidised by long exposure to the elements.

Our guide stopped as we neared them, and in traditional custom, called out to the land, introducing himself and his guests.

We waited, silently, as Sarah and I had first done in Harold's company. With a broad grin, he soon indicated we'd been welcomed, and motioned us on to stand at the foot of the rocks.

At first they seemed a jumble, but on closer inspection, chiselled patterns and images began emerging from rock surfaces. There were animals, birds, fish, human figures, and seemingly abstract designs. Anatomically accurate, many of the animals were recognisable. On one rock surface was the outline of a Thylacine, its body stripes and stiff tail clearly depicted. Thylacines had become extinct in this region many millennia ago, surviving in southern Australia only to be finally hunted out by European settlers in Tasmania. The last of its kind died alone in a Hobart zoo in 1936. Yet here its ghost was to be seen in the landscape it had once roamed.

Our guide also pointed out what looked to be a kangaroo with a plump tail. It was a Sthenurine, a megafauna relative of modern kangaroos that likely walked with a slow gait rather than hopping. "He would have been good eating, that one," commented our guide with some enthusiasm, and I could imagine the nutritional stores of this animal's robust tail being a prized food source. These fat-tailed kangaroos had become extinct by the end of the Pleistocene, some 12,000 years ago.

Looking as though they had been chiselled only yesterday, this ceremonial art site made the lives of those ancient Aboriginal people seem immediate and tangible. How would they have interpreted the decline of their most important prey animals? What stories did they tell to explain the rising sea levels that flooded their coastal hunting grounds, forcing them into a drying landscape? It must have been a harrowing and inexplicable series of circumstances.

It seems to me that the pertinent question is not how they survived, but how did they *learn* to survive?

It is worth reminding ourselves just how significant the sounds of the natural world would have been to the senses of ancestral peoples. Wherever they lived, their soundworld would have been native to them throughout their entire lifespan. It's reasonable to assume they would have known it intimately. As we've found, sound tells us what is happening. So we can be confident that listening would have been one of the primary sources ancestral peoples had for interpreting the living world of which they were so inextricably a part. What did these 'Listening Peoples' learn from their environment about how to live?

The songs and calls of Australia's birdlife speak of co-operative living, social interaction and life-long pair bonding. I imagine ancestral Australians hearing these interactions, noting their associated ways of life, and learning from them. Where conditions were favourable, both birds and people settled and became resident, establishing communities around the abundance of wetlands, rivers, woodlands and floodplains. Through song, birds negotiated their relationships and living space, while people gathered around campfires to sing of identity and belonging to the land. Elsewhere, birds and people moved to follow opportunities. Inland groups listened out for nomadic honeyeaters and waterbirds, tracking them between waterbodies in response to rainfall and good seasons. The calls of nyi-nyis coming in to drink could have led them to ephemeral waterholes. The peoples of the southeast would have heard the cries of currawongs calling them up into the high country over summer to feast on the seasonal bounty. And all across the continent, Aboriginal language groups co-existed with birdsong dialects, while both people and butcherbirds established songlines that told of relationship to place.

This traditional knowledge is remembered today. I recall a talk by Mary Graham, an Aboriginal Kombumerri person and academic, in which she stated quite straightforwardly that when the ancestors of today's Aboriginal People first migrated to Australia all those tens of thousands of years ago, they weren't human. In her gentle, matter of fact way, she went on to explain that when they arrived, they were akin to animals. They only became human here, on the Australian continent, because it was the land that gave them culture. It was nature that made them human.[46]

Viewing the rock art at Murujuga, it was evident that the natural world formed the singular focus of those ancestral peoples.

It seems reasonable to me that Mary Graham's statement can be understood as that Aboriginal Peoples developed a culture sensitive to Australia's ecology. Whatever role they may have played in the late Pleistocene demise of megafauna such as the Sthenurines, they eventually came to adopt forms of prey husbandry appropriate to the country, and learned to manage their own impacts sustainably. They listened to the land and the ways of birds and animals around them. They found out how to exist within the bounds of ecological systems. Rather than conquer and tame the land, they adapted to it.

For me, this helps explain the continuity of Aboriginal culture – it has been grounded in the stability of nature itself. Western culture, on the other hand, has been built on more cerebral foundations – ideas, especially utopian visions of bettering things.

Western aspiration is reflected in the great irony of Murujuga – it is both a major site of human cultural significance, and sits atop a vast underground gas field. As we finished our tour and retreated from the midday sun, the industrial architecture of refineries and flaring stacks emitting hydrocarbons into the atmosphere were visible nearby. Ancestral Aboriginal peoples had survived here by listening and learning from the land, and in doing so, created humankind's most enduring culture. Murujuga is both testament to that, and representative of a new existential challenge.

The Early Europeans

The bush will tell you things – these words have been with me for many years. From my limited understanding of Aboriginal thought, I sense it as having grown from a deep listening to the natural environment. I'm sure it's not this simple, that the influences that have generated culture over generations are many, varied and possibly unfathomable. Yet I find intriguing parallels between what I understand of Aboriginal ways and those of the singing birds of this continent.

What about the peoples whose cultural ideas have come to so influence the modern world – the Europeans? While Australia's Aboriginal People were surrounded by the sounds of birds engaged in social communication, countersinging, listening and responding to each other, duetting sublimely, interacting in communities and creating song cultures, the early Listening Peoples of the temperate north were hearing something very different.

After long winter months of very little activity, spring brings an influx of life. Birds arrive, establish an area in which to live, and proclaim their presence. Every corner of forest, woodland, glade and riverbank comes alive with song, with individual birds staking out their patch and singing each morning from an optimal location.

These vocalisations are adorned with great sonic variety, and northern peoples would no doubt have appreciated birdsongs as a language of extrovert and virile utterances. They would have observed that it was the males making them, the female being the one he was performing to attract. They would also have noticed that once a female had been wooed and breeding begun, his efforts lost urgency. In this spring activity, birds could be seen to sing independently, with little social interaction. Any winter flocks would by then have dissolved, with no foraging associations replacing them. Once summer had finished, the pair bonds of those birds remaining could be seen to be ephemeral and therefore only for the function of breeding and rearing of young.

It is easy to imagine how these observations could be translated into a set of assumptions about life. It is males that are naturally assertive and controlling. They assume dominance by showing off in competitive displays of skill, aimed primarily at impressing females. The female, in contrast, is relatively silent and submissive, being attracted by the male's physical vigour and control of resources. Status is measured by ownership of territory. Success is a virtue of the individual, rather than the collective.

Seasonal contrasts imposed another rationale on northern peoples, particularly once farming had emerged in the Neolithic period. Rather than sustainability, survival was dependent on the acquisition of surplus. Through gathering and storing resources during the warmer months, lean times over winter could be endured. This imperative necessitated the exploitation of seasonal bounty to its fullest extent.

Putting these elements together creates a picture of peoples whose beliefs have been shaped by their understanding of the natural world around them. The showing off and competitive status seeking of males, the valuing and ownership of land and a celebration of individual achievement are all foundational Western assumptions about what is good and natural. In our material economy, the imperative of surplus is echoed in the priorities of profit.

These core values can be seen to have come down to us across the ages. Judging by a long history of warfare over territory by male-dominated societies, they've endured from early times – a seabed over which currents of cultural identity and belief have flowed over many, many centuries. They can be recognised in the thinking of classical antiquity, from which much of our Western world view has been inherited. The science that replaced Medieval mysticism was indeed a revolution, but in means, not foundations. Presumptions about the human dominion over nature remained unquestioned. The universe became conceived of as a mechanism, like a giant astrolabe, the movements of Earth and planets regulated by knowable forces. If the Earth is a machine, and the forces that govern it understandable, then why not set them to work for our benefit? This thinking opened the way for the Industrial Revolution.

Knowable processes became the focus for enquiry into the living world too. Throughout the twentieth century, animal behaviour was viewed as largely deterministic, as though creatures were wind-up toys, driven by biological needs according to the dictates of their genetic code. Meanwhile, ecological studies focused on quantifying the physical dynamics of living systems. More recently, we've become far more open to seeing animals as emotional, having minds and agency of their own – although perhaps this is an expression of our psychological age.

Our contemporary understandings of nature continue to be coloured by cultural assumptions. It is no accident that the language we've inherited from eighteenth and nineteenth century biological studies reflects the world view of that time. Colonist species establish new ecosystems. Animals hold territories. They defend them aggressively against rivals. They compete for dominance. Creatures struggle to survive. Species gain advantage. There are winners and losers. Natural selection progresses toward 'higher' animals, often through an evolutionary arms race. The 'fittest' survive. Nature is a battleground, while places of human battle are referred to as 'fields'.

Language and its unrecognised associations trip us up. Our words often end up saying more about us than the natural world. Acknowledging this is a necessary step toward deeper understanding, as it is through language that we project ourselves onto the tabula rasa of the wild.

All this suggests that at the primal roots of European thought is a fundamental misinterpretation of nature. The behavioural limitations of northern songbirds have come to be reflected in the conceptual limitations of European thinking. This is understandable, perhaps even inevitable, given the restricted biogeographical context of early peoples. From a wider perspective though, we can understand that the extrovert singing of northern birds has evolved not to outdo rivals, but initiate breeding within a brief seasonal window. It is the female who chooses her virtuosic mate, and thus shapes inheritance and future generations. Northern birdsong is more individualistic due to the northern climate.

It may seem incredible – perhaps unbelievable – that our modern world view could have eventuated from such innocent misinterpretations of the natural world. Yet, as the saying goes, from tiny seeds big trees grow, particularly given millennia for ideas to be passed down, expanded, affirmed and by increasing degrees, become unquestioned from one generation to the next.[47]

In our contemporary era, we seem to have arrived at a culmination of these core assumptions. We assume human primacy and the right to control nature as a given, with management strategies applied to exploit practically every resource on Earth, while genetic modification technologies are being developed to alter the very blueprint of life. Our human impacts on the planet have even led to the proposal of naming a new geological epoch after ourselves, the Anthropocene.

Reflecting on all this, I'd characterise the culture that emerged from northern lands, and in which I've grown up, as possessing a self-importance and unhealthy seriousness to it. I wonder whether this has resulted from another circumstance of nature – that northern peoples never had kookaburras to laugh at them, cockatoos and galahs to show them how to play outrageously, and choughs to remind them of the essential ridiculousness of life.

Echoes of Wild Listening

The soundscapes of wild places no longer imprint themselves on our modern brains. Instead it is technology and the pace of contemporary communication that shapes our minds. This makes it difficult for us to imagine the pervasive influence that nature's soundworld would have had on our listening forebears, or even believe that it could be so profound. Yet the reverberation can be detected in the cultural foundations we have inherited.

The origins of these foundations are lost to prehistory, to peoples who listened to and were intimate with nature. Like the footings of a building, subsequent thought rests on the way they saw their world. By the time some cultures came to write down stories, the assumptions they held were already fully established.

By listening to nature ourselves, we can get some inkling of what influenced those Listening Peoples. We can hear that the ecologies of Australia and Europe in particular, are both geographically and biologically on opposite sides of the planet. They represent extremes.

Is it any wonder then, that when the two cultures originating in those bioregions finally met, after tens of thousands of years of independent development, it was possibly the most profound collision of ways of living in human history? The consequences continue today, with the damage done to both Aboriginal communities and Australia's natural environment representing a gulf of ideas between two cultures from two ends of the Earth.

It seems to me that deep listening, both between peoples and to nature, is essential to resolving the issues this has precipitated. In the final chapter of this book, I will explore the practical applications of deep listening in our contemporary world.

For now though, I find it astonishing to consider that the sounds of nature, and birdsong in particular, may have so profoundly influenced the development of cultures. Perhaps this should not be surprising – in nature everything is interconnected. Indeed, if our ancestral patterns of thought had not been influenced by the great communications of the biosphere, we'd have to explain why we should be so uniquely disconnected and oblivious.

Listening allows us to appreciate, in sometimes surprising and unexpected ways, that any separation between ourselves and nature is illusory. It can bring us a little closer to ambiguous truths. Like a Zen koan, the natural world is non-linear and eludes our intellect. Vibrantly alive, present to our senses, instinctive and intuitive, obvious yet baffling, purposeful yet mysterious, nature is as enigmatic as the act of listening itself.

Our cultural narratives, which have their genesis in oral traditions passed down from Listening Peoples, are a tangible yet unacknowledged link that places our most fundamental patterns of thought within the tapestry of the natural world.

Chapter 12

The Communicating Biosphere

The Ages of Sound

The fossil record does not preserve sound. Nevertheless, we can confidently assume that the Earth has been alive with the acoustic signals of living things for a long time.[48]

The first vibrational communications may have been transmitted through inshore marine waters of the world's primitive oceans. At this early stage, specialised acoustic receptor organs may not have developed, nor been necessary, as the dense aquatic medium would have allowed vibrations to be perceived directly by an animal's body. Once life emerged on land however, terrestrial arthropods and tetrapods would have found the far thinner atmosphere incapable of transmitting vibrations as efficiently. Consequently, the earliest terrestrial communications may have been made by drumming or tapping, sending tiny vibrations through stems, roots, leaves or the soil matrix where they could similarly be felt rather than heard. A surprising number of modern insects continue to communicate via substrate-borne vibrations to this day.[49]

Some insects did eventually evolve means of both creating and detecting airborne communications, and some two hundred million years later, crickets, grasshoppers and katydids can still be heard chirruping among grasslands and forests. Somewhat more recently, by the late Cretaceous, cicadas had evolved the capacity to create their loud, buzzing songs, tuning them to daily temperatures, light levels and the seasons.

By this time, tetrapods and amphibians had developed the tympanic acoustic organs from which our ears are ultimately descended, and probably begun their first forays into vocal communication. Frogs may have been among the vanguard, calling discreetly from soaks and wetlands while dinosaurs waded in the shallows.

It is not known whether dinosaurs would have roared as we may imagine. If they did vocalise, it may have been like modern crocodiles, employing the larynx. It's also been speculated that some types of dinosaurs could have funnelled air through extended nasal passages and hollow spaces in their skulls, resonating sounds similar to the way we may blow a conch shell. Another possibility is that the closed-mouth vocalisations of birds had a precedent in dinosaurs 'booming' with a flexible oesophagus. 🐦 It even remains conceivable that some Theropod dinosaurs had a rudimentary syrinx, and communicated in ways that heralded the birds they would become. At some point along the way, mammals developed the expressive possibilities of the larynx, an inheritance that has shaped our own species.

Who knows what wonders of animal communication may have existed at one time, to be subsequently lost as the conditions of the world's biosphere altered? Over the Earth's ages, new repertoires have emerged to signify the divergence of new species. Sonic strategies have arisen, with the successful ones being elaborated on to inform a radiation of new species across genera and even families of creatures. Truly significant developments, such as vocal learning, have transformed the communicative world, filling the air with birdsong and the oceans with the songs of the great whales.

We exist in a moment of this unfathomable Earth time. I find it humbling to consider that everything we hear when we go out into nature now is the result of unimaginable aeons of life experimenting, refining and discarding. Thus I no longer envisage the tree of life as static, but dynamic with the flowing sap of communicative interactions. Novel acoustic behaviours have been the buds that have emerged to inform the growth of new branches. Sound has shaped the living world we now inhabit. Invisible, intangible, immaterial and ephemeral, sound leaves no mark. Yet it has played a crucial role, fashioning the lives of animals, supporting their daily existence, and eventuating in the wonders of a rather noisy biosphere.

Sound has been the dark energy of evolution.

Why Biodiversity?

The first international flight I took in my life commenced a six-month backpacking trek through southeast Asia and India, concluding in London. I was twenty years old, and it was a life-changing journey that would open me to the world, preparing me as a traveller for the field recording trips that Sarah and I would undertake later.

From Sydney, the plane ascended over the forests and gorges of the Blue Mountains, and headed northwest. Gradually, woodlands and agricultural country gave way to the deserts of outback Australia, and I sat transfixed for hours as a seemingly endless vista of orange dune-fields, dull olive scrublands and occasional crystalline salt lakes passed below. Hours later we crossed a remote mangrove-fringed coastline, with coral reefs and sky blue coastal waters, before passing on over a deep indigo ocean. Eventually the plane descended on the emerald green landscape of tropical Bali.

This was my first tangible experience of how small our planet really is. In those hours, I'd traversed a range of ecosystems, from temperate forests and woodlands to deserts, estuarine mangroves, reefs, expanses of ocean and tropical rainforests.

Considering the finite extent of the biosphere, how has nature evolved this level of biodiversity? What drives nature to do so?

This is not simply a question for scientists, a matter of the mechanisms behind evolutionary biology. At its heart, the question is one of the purposes of life. As we are living beings, it is a question that each of us answers for ourselves, in one way or another. This explains why, despite biology being a subject only superficially taught during my schooling, the question was addressed directly. Of all the things I remember from my formative years, one message stood out: 'everything you need to know about life you'll learn on the football field'.

I recall this so clearly because I hated football. Getting ground into the mud by a heathen hoard while trying to get a ball over a line seemed a miserable way of spending a weekend. I'd do anything to get out of playing. This prompted my mother to ask our headmaster if there was an alternative, his dismissive and frankly insulting response being something she never forgot. But it did represent the quasi-religious fervour with which football was held. Through football I would learn to play the game – the game of life.

So, what is the game of life?

Implicit in the football metaphor is that life is competitive. Competition – we've been given to believe – is universal, and always results in the most optimal of outcomes. Competition challenges us. It punishes laxness, incompetence, laziness and inefficiency. It encourages excellence. It brings out the best in us. It is necessary and essential.

And it is natural. Given the significance of competition in our social world, is the assumption that it fundamentally shapes the natural world reasonable? I'd like to pose a scenario to consider.

Imagine a human competitive interaction such as a sporting tournament. Once the competition has begun, there are no new participants. At first, there are many contenders, but successively, they are eliminated until the competition ends with only one, single, winning individual or team. They are the ones holding up the cup, receiving the gold medal, or spraying the champagne.

You can replicate this in any human system based on a competitive process – political, legal, commercial, professional.[50] The point is, in all of them there will be more participants in the beginning than at the end. In our social model of competition, diversity diminishes as the process proceeds.

If nature were driven by a similarly competitive process, a 'winner takes all', 'survival of the fittest' mode of functioning, then why does biological variety *increase*, rather than wither? Why do habitats establish and through a process of ecological succession, come to support an ever greater diversity of species? Why have living systems not devolved to a few ultimately supreme creatures? It's not as though life hasn't had enough evolutionary time for a handful of planetary winners to inherit the biosphere.

And yet, it hasn't happened (notwithstanding our tacit belief that humans *are* that species). Wherever I've travelled in the world as a nature sound recordist, in every wild environment that remains in a reasonably intact state, my microphones document the sonic expression of biodiversity. In each place – an Indian forest, under a baobab tree in Africa, in a European woodland, or my own bushland at home – natural soundscapes are rich, complex and quite unique.

Amidst all this variety, I have nevertheless come to recognise a coherence uniting nature's soundworld. I hear each local manifestation as an expression of something far vaster: the great communications of the biosphere. It is as though nature, while expressing itself in a glorious variety of songs and calls, nevertheless speaks with a universal voice. This voice is one beyond species and their behaviours. It is more than the sum of its parts. It transcends time and place, seasons and habitats. It is a voice that speaks of the underlying processes by which the wonders of biodiversity are brought forth and maintained.

If nature can be thought of as the game of sustaining life, then by listening, we can hear its rules of play.

The Origins of Sonic Communication

Competition can certainly bring animals into conflict. So too can predation. Yet they're very different interactions. Predation concerns one animal utilising another (almost always of another species) as a nutritional resource. Competition involves organisms (of the same species or not) that seek to utilise the same resource.

Predation is a central process to our Earth's biosphere, and the relationships between predators and prey have long been thought to drive evolution. The prevalence of that assumption has been tempered more recently by other ideas, something I'll come to shortly. For now though, there is another outcome of predation more pertinent to us as listeners; it has indirectly led to sonic communication.

Across the animal world, many creatures react with a 'startle response' when disturbed or attacked. This often involves making a sound or vibration to confuse or intimidate a predator.

Among arthropods, stridulation is a means by which they create this defensive response, achieved by the scraping of two rigid body parts; usually a scraper or 'plectrum' against a series of hard ridges. Depending on the type of animal, these ridges and scrapers may be situated on wings, legs, edges of mouth or abdomen – basically anywhere they can be rubbed together to generate a sharp buzzing vibration. Think of dragging your fingernail along the teeth of a hair-comb. For a predator, this startle response must be like having an electric buzzer going off in their face. 🐦 Even predators without hearing could be deterred by this strong and unexpected vibration alone.

Which is the way the fossil evidence points to this defence first evolving. Stridulatory surfaces have been found on early arthropods and invertebrates dating back to times before auditory sensory organs had developed, suggesting them as functioning to generate this kind of protective response, rather than for any communicative function.

Ancient startle responses have significance for us because modern insects use similar stridulatory mechanisms to communicate. While many varieties of terrestrial insects have been found to stridulate, they often do so quietly. The louder types more familiar to our ears make their calls by rubbing a plectrum on one wing against ridges on another (in the case of crickets, mole crickets and katydids) 🐦 or against legs (grasshoppers). In aquatic environments, a fauna of boatmen, backswimmers and water beetles similarly use stridulation to create the buzzes, zizzing and clicks of underwater soundscapes. 🐦🐦

So while the evolutionary development of stridulatory mechanisms in various lineages of insects is complex, their presence early in the fossil record does suggest that they first evolved for defensive purposes, and only later have been adapted to communication.

The other notable group of terrestrial soniferous insects, the cicadas, suggest a related and rather fascinating history. Cicadas produce sound by flexing a hard chitinous membrane on their abdomen, like a drumhead, the resulting vibration being modulated by muscles around the membrane and amplified by a hollow abdominal structure. The volume they can create make cicadas the loudest terrestrial animal for their body weight.

The earliest cicadas, the Tettigarcta, also known as hairy cicadas, have their origins in the Triassic period. They became ubiquitous by the late Mesosoic, occupying habitats from the tropical to temperate. During this time they had neither ears nor the membranes for sound making.

What was the evolutionary impetus through which modern cicadas, which evolved after the Cretaceous, developed ears and acoustic membranes when they'd not done so previously? My suggestion is that birds, which evolved and diversified around that time, became major new predators for cicadas. In response, cicadas developed their own form of vibrational startle response. This amplified over the course of evolution into a novel defence strategy that many larger cicadas still use today: a sonic wall of deterrence. A tree shimmering with calling cicadas has not only been observed to deter birds, but makes locating individual insects difficult. In this way, cicadas have evolved their own version of the same trick that the anurans developed.

Cicadas will also give loud, buzzing startle responses when attacked. Watching our honeyeaters chase them on the wing, only to drop them upon being 'buzzed', is very amusing. Not that this works every time. Bee-eaters are skilled aerial hunters and very fond of cicadas, migrating all the way from tropical Australia to our southern bushlands to breed and feast on them during the summer. So cicadas don't get it all their own way, probably remaining a rich food source for many breeding birds over warmer months.

In this scenario, cicadas initially evolved a means of producing sound to defend themselves from predation – perhaps not completely, but enough to be a worthwhile strategy. Only later did they develop organs for detecting sound, and extend their vibrations into courtship and love songs.

Meanwhile, from the wealth of early hairy cicadas, only two relic species have made it into the modern era. Both are found in Australia, one in Tasmania and the other in the high country of the mainland southeast. They are semi-nocturnal, only coming out at dusk, spending the day sheltering under strips of bark. They neither call acoustically nor have auditory perception, their only defence against the predation of birds being to hide during the day.

All this points to the mechanisms of emergency defence being subsequently refined for nuanced signalling among acoustic insects. It has likely been so for early tetrapods and proto-amphibians too. The fossil evidence is unclear on the timing and circumstances under which tympanic ears evolved, but predation, either the finding of prey or for warning against threats, is likely to have driven their evolution. While some modern amphibians and reptiles remain largely mute, the larynx, vocal cords and syrinx have developed in others, eventuating in the wealth and diversity of animal communications we hear today.

Once communication had been established, predation has continued to shape how it is achieved. We've met our Common Bronzewing, and heard its calls as a form of acoustic camouflage. Alarm calls are another development, with signals often being shared among species – honeyeaters and drongos being well recognised for giving a warning that other birds respond to by seeking shelter.

One sophisticated strategy, widely found in the animal world, is the opposite of diving for cover when a predator is around. In this behaviour, an animal goes out of its way to make itself obvious. This may not seem a clever idea, however the creatures that do so usually have a 'secret weapon' defensive strategy, such as being poisonous, prickly, spitting toxins, or emanating a particularly unpleasant odour. As effective as these secret weapons may be, they are of limited value if a predator only discovers them in the process of attacking. Hence a display to forewarn is required.

These are termed aposematic displays, and when threatened, visual signals often involve making the body appear larger, possibly by arching the back, standing on tip toe, raising hackles, fur or feathers, opening the beak wide or baring fangs. A cat in defensive pose is a classic aposematic display. It means; 'don't even think of trying it, this won't end well for you'. A bear raising itself to stand on hind legs and the vivid colours of poison dart frogs are advertising the same thing, as are the bold fur markings and nonchalant air of the skunk.[51]

Aposematic displays are often accompanied by specific vocalisations. The low growl of a dog, or the rattle of a rattlesnake, are audible signals to back off or else. Hissing is an almost universal warning, heard in a range of animals from lizards to felines. Most modern frogs make anti-predator calls when disturbed, a trait that likely dates to their origins in the Jurassic period.

Protective displays of this kind are instinctive and primal, and there's little reason to believe they don't have deep evolutionary roots. Although we may never know for sure, it seems reasonable to imagine that defensive throaty hissing from proto-amphibians and early reptiles has been the evolutionary precursor to vocal cord vibration.

All these narratives tell of predation having generated a variety of acoustic defences. For some animals, over the vastness of evolutionary time, these have transcended their defensive purposes to ultimately become the communicative behaviours we hear around us today; frog choruses, mammalian calls, birdsong and choirs of crickets on a summer evening.

So while predation has been used to present an image of nature 'red in tooth and claw', we can instead view it as having shaped our world positively, promoting sophisticated behaviours, nuanced communication, and in the variety of nature's soundscapes – transcendent beauty.

Dancing around Competition

From our definition – creatures seeking to utilise the same resource – the potential for competitive conflict obviously exists. Food resources, nesting hollows, homeranges, mating opportunities; all these are finite resources that may bring animals into contention. For mammals, vying over status and access to breeding females can bring males into very physical conflict. It is the classic scenario of two creatures contending over the same resource, and these male-on-male conflicts are often the documentary trailers advertising a competitive natural world.

The issue is not whether contentious circumstances can arise, but how creatures behave when they do.

Of all animals, I can't imagine a species with more potential for dramatic fighting than the Hippopotamus. When we had the opportunity to witness them doing so, it was at somewhat more close quarters than we were comfortable with.

Once the idea of a field trip to Africa had taken hold, Sarah and I decided to undertake it very differently than the usual week-long, luxury-in-the-bush, packaged tourist safari. Scheduling two months in Tanzania, we hired a Land Rover with the expectation of camping for the duration, plus opting to employ a local driver and cook. Fortune smiled on us with Roger and Ally, who were both highly capable and of boundless good humour. Despite being a self-trained naturalist, Roger turned out to be one of the most knowledgeable guides we've ever worked with, and the four of us bonded immediately as a little travelling party.

In the remote southwest of Tanzania, Roger drove us into a grassy clearing, an informal campsite on the banks of the Katavi River. Leaving Ally to set up his kitchen under a grass thatch gazebo, Roger, Sarah and I walked a hundred metres down to the riverbank. There, wallowing in the 'great, green-grey, greasy' waters, were upwards of a hundred hippos. Roger was under-standably concerned, as they have a reputation for being the most dangerous animals in Africa, killing more villagers than any of the big cats. They're unpredictable, sneaky, and have a reputation for attacking without provocation. Despite seeming ungainly, they can move very quickly when they want to.

Watching these car-sized animals from the riverbank, and occasionally being given a full-gullet view of their mouths and fearsome incisors as they yawned, had us appreciating Roger's concern. However there was a wooden fence around our compound, and Roger considered we'd be safe enough if we pitched our tents close to the vehicle.

An hour later, with dinner finished and the light fading, Roger got up to walk back to the Land Rover. His "Uh oh!" had us looking over to find a hippo nonchalantly grazing in the middle of our campground within a few steps of the vehicle. "I should have known" he said, pointing out a fallen fence railing, and what looked like piles of lawnmower clippings piled in drifts against many of the fenceposts. Hippos have an unusual manner of defecation; they back up to a tree or rock, and spinning their stumpy tails, spray desiccated vegetation like a firehose as a means of marking their patch. Looking around, we realised there was nowhere that wasn't 'their patch'.

That night, with our tents pitched protectively close to the vehicle, I put microphones out on its roof to record the nocturnal activity. Half an hour after bedding down, I heard the breathy, half panting, half growling of a Leopard not far away. Oh great – we had both a sneaky herbivore, and now a sly predator patrolling nearby. I consoled my anxiety by realising I'd captured a recording of it.

In the middle of the night we were awoken by a great bellowing and roaring. Two hippos were going for it, somewhere beyond the edge of the camp. ♪ For half an hour we listened uneasily for any sign of them getting closer, but eventually they fell silent. The following morning, the hippos were all jostled together flank by flank in the water, happy families once again. ♪

Several days later, we witnessed another fight, this time by daylight. With mouths open, two hippos were face to face, pushing each other, tusks locked in what looked to be mortal battle. Once again, the roaring was thunderous, and we withdrew to safety, watching on as the huge animals wrestled, pushing each other back and forth. Eventually one broke off and trotted away. For such powerful and weaponised animals, I was surprised to see their injuries appeared to be only a few superficial scratches.

What struck me about the interaction of these male hippos was a sense of formality about their behaviour. Rather than simply attacking each other haphazardly, they seemed to have rules of engagement. Walking together, they'd almost gently lock tusks before beginning the tussle. They also seemed to know when the encounter was concluded, and despite it not being at all obvious to us what the outcome was, they knew.

That hippo contests are more ritualised behaviours than 'no holds barred' fights, was an interpretation reinforced shortly after when we observed a similar interaction among giraffes. In an open area, a pair of males strode toward each other until they were standing nearly alongside. Pulling their necks away, they swung them back forcefully, striking them together like crossed sabres, connecting with a fearful thump. Stepping back slightly, they gathered themselves for another swing. This required co-ordination, as they had to synchronise their movements. If they didn't swing together or connect properly, the considerable momentum involved could easily result in overbalancing and a fall, with a broken leg having fatal consequences. So they danced, kicking up the dust, whacking their graceful necks together. It was a strangely elegant performance, both unwieldy and powerfully physical, which again, seemed to resolve with one animal staging a dignified withdrawal.

Observing the interactions of iconic animals in Africa was intriguing, but somehow, seeing a similarly formalised behaviour between the wallabies living in our bushland really brought home to me how universal these rituals of conflict resolution are. Wallabies have a shy and solitary temperament, so I was surprised one morning to come upon a pair of males vigorously shirtfronting each other. Standing tall on hind legs and tail, they were chest to chest, their forearms around each other, heads craned back out of each other's reach. Thus entwined, they were pushing back and forth, each trying energetically to knock the other off balance. Often leaning back on tails, they were bringing their powerful hind legs up to land a kick on the other. Occasionally one would stumble and go flying, tail and limbs akimbo. Springing back up, they'd rejoin and continue.

There was something about their tussle that reminded me of sumo wrestlers attempting to upend their opponent. Sumo is a highly disciplined and formal sport, with very strict protocols. Watching these wallabies I could see it was not a free-for-all, but involved a similar discipline. They were engaging by a set of unseen rules, without which the interaction could not have functioned.

After a minute or so they disengaged and moved off in different directions, panting heavily. I thought that was the end of it, but no. Breath regained, like boxers in a ring, they drew together for bout two. This had them pushing each other between the trees, quite oblivious of me, trying kicks or whole body twists in an effort to best the other. Eventually they disengaged and one moved away, not to return. The other, presumably having proved itself the dominant animal, was squatted, body low to the ground, breathing heavily. He looked utterly spent. When he soon hopped off I could detect no sign of injury, either on his body or in his movement.

All these behaviours are termed by biologists 'agonistic', their purpose being to moderate the physical costs of conflict. This is in contrast to *ant*-agonistic actions, which inflame and escalate a situation. Antagonistic interactions between animals are exceedingly rare in nature, whereas agonistic behaviours are well-recognised and ubiquitous. Indeed, most species seem to have their own rituals by which they diffuse and resolve conflicts. These may include displays, bluffs, conciliation, placation, submission, retreat or, as I'd witnessed, formalised expressions of aggression.

That animals would evolve ritualised behaviours for resolving contention makes sense. Physical conflict risks significant injury. Minimising the chances of this happening to healthy males increases the viability of the whole population. While a subordinate male may not be successful in one encounter, he will survive to possibly take up an active breeding role in the future when maturity or circumstances allow. Damaging genetic diversity through mortal fighting is not advantageous to the survival of a species. Natural selection can be expected to ensure that animals whose behaviour minimises this possibility have an advantage.

I'm not suggesting that aggression, and even mortality resulting from it, never happens. Sometimes it may. But overall, evolution is a diligent accountant, and weighs any advantages of belligerence against the risks. Hence animals have inherited elaborate species-specific behaviours to mediate their relationships without jeopardising their own wellbeing, or causing serious injury to their own kind. This would explain why agonistic rituals have evolved to be so pervasive. I guess life is unpredictable enough without adding to the dangers.

Physical aggression may also be inefficient, expending valuable energy for only short-term benefit. The honeyeaters that feed in the correa bushes at the front of our home gave me a demonstration of just how temporary that benefit could be. One morning I was idly observing an Eastern Spinebill as it probed flowers in the middle of the bush, hanging upside down and assiduously working its way from one blossom to the next. Meanwhile, a Yellow-tufted Honeyeater, a slightly larger bird, watched from a tree nearby. Suddenly it darted down into the correa, the two birds exploding out the far side, the yellow-tuftie hot on the tail of the spinebill and snapping its bill loudly. The two climbed up, up and out of sight over the top of the house, at which point the yellow-tuftie broke off pursuit and, rather than going to the correa, simply returned to the tree it had been in. "Well, I hope you're pleased with yourself", I commented at its seemingly unnecessary bullying. Seconds later, a purr of feathers whizzed past me as the spinebill, having flown over the house, under the carport and round the corner, shot straight back into the correa to continue feeding. The yellow-tuftie watched but didn't react, flying off shortly thereafter.

It is easy to view this kind of behaviour as aggressive, a way of asserting rights to resources. However I've found that it's worth pausing to reconsider this assumption. I often see birds chasing each other; sometimes they appear to be a mated pair flirting, or members of a social group animatedly pursuing each other around the tree canopy. A reasonably familiar sight is several young magpies scrapping on the ground in what seems a confusion of feathers and squawking, but which in reality has social bonding purposes.

On one occasion I witnessed two family groups of choughs, each comprising around a dozen birds, flap down onto a patch of open ground and strut purposefully toward each other like opposing armies. The ensuing melee had birds chasing each other, feathers splayed and filling the air with harsh calls. I even saw birds flat out on their backs, wings spread, with several others standing on top of them. It was Waterloo. After a few minutes, as suddenly as it began, the two groups disengaged without visible injury, shuffled their feathers back in place and flew off in opposite directions.

Although it was too chaotic for me to be positive, I suspect I witnessed a ritual of mock conflict, in the midst of which a young bird was distracted from one family and taken off to be included in another. This is a necessary aspect of the communal living of choughs, their way of exchanging genetic diversity between groups. Rather than a kidnapping, it can be viewed as not that dissimilar to the way human tribes and societies intermarry.

In all these cases, what we perceive as aggression may have other interpretations. Even the notorious belligerence of our honeyeaters can be seen in another light. When resources are abundant, multiple honeyeater species co-exist with minimal contention. It is only when those resources dwindle that the larger species – those that require rich concentrations of nourishment to survive – physically signal to smaller and more adaptable species that it is time to move on. Are they enforcing a pecking order to 'protect their patch', or sharing it in times of abundance and thus supporting biodiversity in a dynamic ecosystem?

For our yellow-tuftie and spinebill, it may indeed have been a brief although ineffectual aggressive outburst, but possibly it was simply an instance of two nectarivorous birds pumped on sugar and engaged in a game of chasing.

Singing around Competition

Animal behaviours show how the potentially damaging costs of competition are minimised through ritualised interactions. Can we hear sound being used to achieve the same purpose?

Staying with mammals for a moment, if any Australian animal suffered from a quarrelsome temperament, it would have to be the Tasmanian Devil. They are concentrated balls of muscle and attitude, and group feeding events at carcasses (they're scavengers primarily) can result in faceoffs accompanied by squeals and screams, an example of formalised aggressive behaviour for agonistic purposes. They've recently been found to have another vocalisation, termed 'arffs'; a quiet, short, low frequency sound with complex characteristics. Arffs are heard most often as animals tug together on a large food item to reduce it to more manageable proportions. The vocalisation appears to communicate individual identity, allowing the animals to put aside their touchy tendencies and engage in a mutually beneficial feeding strategy.[52] So for devils, specific sounds allow them to both minimise harm in physical altercations and facilitate co-operation.

Looking beyond mammals, I think it's very possible that much of what is interpreted as sonic signalling to attract mates or repel rivals has an equally profound role in reducing conflict. The a-synchronised calling of frogs seems likely to fulfil that function, and insects exhibit similar signalling patterns. But of all creatures, it is birds that have finessed the use of sonic communication.

We should expect that, like mammals, they negotiate their relationships in a manner that minimises harm. Unlike mammals though, they seem to have evolved a means of doing so that is entirely non-physical. Instead of fisticuffs, they sing.

Earlier, I spoke of the dawn chorus which, while it can seem a riot of birdsong, is no anarchic, free festival, but composed of highly refined behaviours. I've thus suggested that it likely serves a negotiatory purpose. This is indicated by the many vocal behaviours heard in the dawn chorus which are unique to that time of day.

White-eared Honeyeaters, our 'kWHI-choo' birds, embody several aspects of this. Throughout the day, we hear a variety of different calls in addition to *kWHI-choo*; white-ears also give loud, chopping syllables or a soft, bubbling trill. These are exchanged across the landscape, as one bird calls in response to another, giving an update on their whereabouts. 🐦 This is a brief reciprocal exchange, not the countersinging of the bird's dawn chorus, which is a finely executed and sustained performance.

Although white-ears don't have a unique dawnsong repertoire as some other birds do, they nevertheless exclusively use a single call. From their range of diurnal repertoire, they only employ their *kWHI-choo* vocalisation at dawn. Because it displays regional dialects, this call is associated with local breeding populations. Interestingly, white-ears sing at daybreak pretty much throughout the year, even when other birds have fallen silent in winter. During those cooler months however, their singing is relatively abbreviated, devoid of countersinging, and done with any of the other calls in their repertoire *except kWHI-choo*. 🐦

Such sophisticated singing behaviours stand alongside non-passerine bird groups – various fowl, cuckoos, pigeons and doves in particular – that may frequently be heard at dawn, but do so in a manner indistinguishable from any other time of day. While they may be achieving similar ends though less sophisticated means, this only highlights how remarkable songbird behaviours have evolved to become.

Collectively, all these vocal behaviours give the dawn chorus its structure and integrity, a cohesion that can fall apart when populations are diminished or species absent, as evidenced by New Zealand's nectarivores.

This suggests to me that birds, and especially the songbirds, have evolved the dawn chorus to negotiate their existence with others, particularly their own kind. The predawn hour is their time of daily ritual in which they perform this significant function; acknowledging each other, spacing their homeranges, reaffirming communities and defining relationships. By the time the sun peeks over the horizon, this avian ritual is mostly concluded. While it may facilitate purposes such as advertising for mates, its most important reason for being is a diplomatic one. The refined behaviours we've noted – countersinging, regional dialects, unique repertoires, sequencing, song mirroring – are the protocols which allow songbirds considerable nuance in how they live together.

By singing at daybreak in such sophisticated ways, songbirds achieve a greater efficiency of living than physical confrontation ever could. For me, the dawn chorus is one of the most elegant demonstrations of how nature can avoid competitiveness. This makes it a wonder of the natural world, and one we can enjoy every morning during the appropriate seasons.

Essential Co-operation

We view competitiveness and co-operation as opposite, yet nature tells us they are not equally viable.

While creatures have evolved sophisticated behaviours for minimising the harm of competitiveness, co-operative interactions have flowered to breathtaking degrees. Some of these co-operative relationships are so foundational to life on Earth that it's easy to take them for granted. The functioning of terrestrial ecosystems for instance, is built on the mutually beneficial relationships between pollinators and flowering plants, and between plants, mycorrhizal fungi and bacteria. Like all animals, when we eat, we are not feeding ourselves so much as our gut microbiome, on which we depend to convert food into nutrients. So while competitiveness seems a hindrance, co-operation and mutually beneficial alliances between organisms are central to the functioning of the living world.

Actually, co-operation is probably not quite the right word. Co-operation for us brings connotations of intention – decision making, agreement, goal setting and so on. There is no such agency for organisms. If the advantages of an interaction outweigh the disadvantages, that may be enough for it to become a preferred way of living.

For ecologists, mutually beneficial relationships come in various shades. Symbiosis occurs when two organisms live in an essential partnership. Other interactions may be more flexible. Mutualism and commensal relationships offer either mutual benefit to both parties, or to only one. Other interactions may facilitate vital aspects of life for either of the participants, or for entire ecosystems.

The detail of these interactions are most often complex. It is easy to view mixed species foraging flocks as a single co-operative behaviour, yet each species contributes and derives benefits in its own way. That drongos can feed almost exclusively within such flocks indicates they are invested in this collective lifestyle. Likewise other species spend much of their day in mixed company and benefit greatly from doing so, while others may only join temporarily as an aggregation passes nearby. So while the potential for benefit exists in a co-operative behaviour, each species may get more, or less, from the opportunity.

This entwining of lives in mutually advantageous ways can be found frequently in nature if we know where to look – and listen.

The sharp, high-pitched whistles of Mistletoebirds tell me when these delightful little birds are around our home. ♪ The male is handsome in iridescent black and white plumage, with a vivid red throat and undertail, while the female is soft grey with a flush of pink under her tail.

Mistletoebirds are named for their close association with mistletoe, both bird and varieties of the parasitic plant being found across the continent. Previously, mistletoes were seen as detrimental to tree health, but more recently have become ecologically appreciated as their flowers and fruit are a rich food source. The hanging foliage of mistletoes is a safe and abundant place to forage, and I frequently observe our honeyeaters, rosellas and babblers piling into the midst of one to rest and feed. As they do so, leaves and detritus fall to the forest floor, creating a microbiome nurturing microbial and invertebrate life in the soil, which then becomes a fertile bed for seed germination. While this assists plant propagation, the hidden recesses of the mistletoe also provide other birds with safe locations to build their nests.

The Mistletoebird feeds largely on the fruit of the plant, which then pass through the bird's digestive tract. When a Mistletoebird defecates, it's not your ordinary bird turd. For a start it is not of the white dollop variety. It is a sticky, mucilaginous dropping containing the undigested seed which, with a neat swiping motion, the Mistletoebird smears onto a convenient branch. From their calling card a new mistletoe may grow. You'll be pleased to know that as they attend to the call of nature, I've heard Mistletoebirds sing.

This mutually beneficial association between bird and plant in our bushland is paralleled by another. The Cherry Ballart, or Native Cherry, is a densely foliaged, cypress-like tree which is semi-parasitic on the roots of eucalypts. Their small, red, berry-like fruits have the highest sugar level in our forest, and call out "eat me!", which our Grey Currawongs in particular are happy to do. Once again, digestive juices go to work, softening the enclosed seed and preparing it for germination. Excretion often happens when the bird is perched in a eucalypt, the seed thus falling near the tree's roots, ready to begin a new cycle.

When I hear the commanding calls of currawongs, 🐦 or the sweet whistles of Mistletoebirds, I know they are both playing their role in dispersing the seeds of ecologically important plants. The plants in turn are creating high nutrient micro-habitats for honeyeaters and other birds, and fertile seed beds for new propagation. Together they are keeping our forest vibrant. Oh, and I have to add that the currawongs are right – the fruits of the Native Cherry are very tasty.

At Ningaloo Reef, off northwest Australia, as the first light crept into the sky, I listened with hydrophones to the barking and drumming of fish. Nature has another dawn chorus. It is to be heard underwater, in marine and estuarine environments the world over. Yes, fish also vocalise to bring in the new day.

Drifting with the current over coral bommies, I could hear the community of vocalising fish congregating around them making quick bursts of staccato grunting. Sometimes these would be occasional, and at others a whole conversation seemed to be going on down there.

This underwater dawn chorus is less exuberant than that of birds, as only certain fish groups vocalise – species such as grunters, croakers and drummers being named for this ability. Soniferous fish make sound in a variety of ways. Some vibrate muscles along their swim bladder, utilising the bladder as a resonator to amplify the sound. Others twang tendons, or scrape and rub hard body parts together; teeth ridges or spines. The swim bladder may then be utilised to modify sounds, adding 'body' to higher frequencies. Whether the sounds made in the piscine dawn chorus perform a similar negotiatory function to the avian one, I've really no idea.

Coral reefs are an ecosystem with exceptional productivity and diversity, which is paradoxical as they exist in often nutrient-poor tropical waters. That coral reefs exist is solely due to a mutually beneficial association between reef forming corals and unicellular algae known as zooxanthellae. Theirs is a multifaceted relationship, in which corals stimulate the algae to release photosynthetically produced organic compounds. Meanwhile the coral provides the algae nutrients and a home within its tissues. What adds complexity to this picture is that corals regulate the algae, controlling not only their secretion of organic compounds but also their population growth and density. Between them both, the corals and algae accumulate large quantities of nutrients, storing them within their living biomass. Without the algae, corals bleach and die.

This fundamental association allows one of the most biodiverse ecosystems on the planet to exist, and as we know from limestone formations around the world, it has functioned for around half a billion years.

However, as I listened to the marine soundworld at Ningaloo, I was hearing something additional to the fish, a sound representing another co-operative partnership that contributes to maintaining reef integrity. Much of the audio in my headphones was coming from small crustaceans variously known as snapping shrimp or pistol shrimp. They are ubiquitous in the world's oceans, such that the snapping sounds they create are referred to as 'marine crackle'. Loud and ever-present, it is the static below the ocean waves.

These shrimp create their 'snaps' with an enlarged and specially adapted pincer, closing it with such force that a high pressure air bubble is formed, the bursting of which results in a loud pop. This generates a shockwave intense enough to temporarily stun nearby small fish – the shrimp's prey. The other pincer is then deployed to casually retrieve dinner. It is an astonishing feeding strategy utilising sound. So the marine crackle familiar to any diver or snorkeller is the sound of a host of tiny shrimp on the hunt.

In addition to being sonic predators, snapping shrimp live with corals in a mutually beneficial partnership. The shrimp protect corals by attacking sea stars that prey on them, in particular the Crown-of-thorns, a large species which can devastate reefs when in plague numbers. An approaching sea star will be met by a feisty shrimp, not only snipping at the star's spines and delicate tube feet with its powerful pincers, but blasting it with shock waves. These defences by the shrimp have been observed to repel sea stars, with experiments by marine biologists demonstrating substantial improvement in the coral's chances of avoiding attack. In return, the coral provide shrimp with shelter among their tightly branched folds, nooks and crannies, and may even offer additional nutrients.[53]

As Ningaloo is a reasonably healthy reef, I suspect much of the shrimp activity I heard was in pursuit of breakfast. During the time Sarah and I were there we went swimming over the reef each day, and didn't notice high sea star numbers. Actually, Crown-of-thorns at normal population levels provide a balancing role on reef communities, preying on faster growing corals while allowing slower ones to establish. In this way they play a positive role, promoting diversity and reef succession.

However human impacts such as overfishing or nutrient runoff from agriculture can result in reef systems getting out of balance, and this is when sea star infestations can have disastrous impacts. If we address the ways we humans create imbalance for the reefs, then the coral and their noisy little defenders should be able to work together to maintain a viable ecosystem.

Mutual Accommodation

There is another form of interaction that lies somewhere between our polarities of competitiveness and co-operation. And it is one that I hear each and every time I'm in nature. For some time I didn't have the words to describe it, but in a conversation with Freya Mathews, Emeritus Professor of Environmental Philosophy at Latrobe University, she casually articulated what I was getting at, and I think the phrase 'mutual accommodation' expresses it well. It is the interaction creatures have when they could be contesting a shared resource, yet instead, they live and let live.

On the plains of Africa, this is readily apparent. Mixed herds of herbivores, often of considerable size, comprising wildebeest, zebra, hartebeest, rhinos, impala and other deer of various kinds, can be observed grazing the savannahs. They all eat the same thing – grass – yet they don't contend with each other over it. Ecologists studying this phenomenon have hypothesised that mixed herds offer additional warning of predators, and hence are advantageous.

Nevertheless, it is a good example of creatures sharing a common resource while living alongside each other.

Mutual accommodation is audible too. Here the potentially contentious resource is the acoustic space itself. Presumably creatures need to be audible to communicate adequately, but are they competing over 'air space' or 'bandwidth' to do so? What I hear in nature makes it evident to me they aren't, and research into animal perception begins to explain why they don't have to.

While we hear the world in a particular way, other creatures have their own auditory and cognitive processes, and may be attuned to quite different aspects of sound than ourselves. Even among humans, we have very different interpretations of what we hear. Frequency perception for instance, varies between those with perfect pitch discrimination, to most of us who can recognise melodies by relative note sequences, while some can't hold a tune to save themselves.

We also comprehend sounds differently according to our cultural backgrounds. The mental processing of linguistic and 'musical' sound is thought to be performed in different cognitive centres, and this is dependent on context. Some languages require a sensitivity to pitch and tonal inflexion that others don't, moving those elements from musical to linguistic significance. The percussive clicks of African Indigenous Khiosan languages have been found to be processed in the musical centres of non-speakers. We may all hear the same sounds, but make sense of them very differently.[54]

In nature, we can expect far greater variety of sensory perception and cognitive processing. For instance, birds have long been thought to be good at pitch perception, as they frequently have such tuneful songs. Recent research on Common Starlings has indicated this is not the case.[55] It turns out starlings are poor at recognising pitch sequences (melodies) and timbre (colour of sound), but have a keen awareness of the spectral envelope of sounds (like syllables and elements). It seems their senses are more tuned to what we'd recognise as a linguistic form of perception than a musical one.

Even what an animal physically hears can't be assumed. The acoustic perception of many insects is tuned primarily to the frequencies of their species' songs, and have low sensitivity outside that specific range. Among birds, Emus have been found to be only able to hear up to about 4kHz (human hearing may reach to 20kHz). This gives them an awareness of their own drumming calls and just enough range to detect the thin, quavering whistles of their offspring, 🐦 but little else. They'd be largely oblivious to birdsong around them.

Incidentally, this may reflect an ancient trait, as the ear canals of fossil Archaeopteryx, that 'first bird' whose fossil imprints date to around 150 mya, indicate a similar auditory capacity. It is likely that Archaeopteryx were also deaf to higher frequencies, and if they vocalised at all, would have, like Emus, called at lower frequencies.[56]

These considerations point to successful communication being an outcome of cognition, rather than physical parameters of sound such as frequency. Even audibility may be a poor measure, as we know from our ability to communicate surprisingly well despite noisy surroundings. By developing unique sounds and refining their perceptual awareness to them, creatures are able to share their acoustic surroundings.

The result is to be heard in natural soundscapes the world over. Not only do creatures share the acoustic space with each other, but they have adapted to overcome the potentially masking sounds of the physical environment: rainfall, the roar of ocean waves, wind in foliage, or flowing water.

Another indicator that acoustic bandwidth is not a prerequisite of communication is that, as often as not, creatures vocalise together – during the dawn chorus, in frog ensembles, as communities of insects on a warm day, during spectacular cicada emergences, while foraging in flocks, at seabird colonies, or when whales congregate on their breeding grounds. Often, nature creates a collective and sometimes intentionally confusing din. At other times, the airspace becomes quite unoccupied, and we encounter that deep quietude that speaks of ecosystem health.

I find it wondrous that the sonic diversity we hear in nature has its parallel in the cognitive perception of creatures, each mind being uniquely attuned to its own communications in ways we possibly cannot imagine.

I also believe this explains the appeal of listening to nature. It's not that all wild voices are intrinsically relaxing, as some may be quite harsh or unexpected. It is that we instinctively respond to nature's sounds as expressions of creatures who share their environment together, a mutual accommodation that pervades the natural world – a world in concord.

This was brought home to me by way of contrast while recording in Africa. At one park we met a wildlife cinematographer who was spending a year gathering material for a one-hour nature documentary. He was out each day, hunting for incidents to film. At first I assumed his brief was to document over successive seasons. However we learned that the film's producers wanted fur-flying footage. The amount of time it was taking him to capture those moments of high drama reflected how infrequent they are. Every day, he would be able to film hours of animals foraging, resting or going about their business. If that was all that was required, filming would be over in a week. But the moments of confrontation that would make the final cut and hit the screen are rare, and even then, may not be the dramas an editor and narrator make them out to be.

Each day, the cinematographer went out, as we did, to pursue field work and gather materials. While he was hoping for even a minute of usable footage, I was gathering hour after hour of sound recordings, all of them richly evocative and a pleasure to listen to.

Nature doesn't readily deliver a dramatic, competitive narrative. Much of the time, creatures live side by side. It is not such a jungle out there after all.

Into Wonderland

It is now time to tell a story. It is, you might say, an evolving tale, having not come to any conclusion, at least not yet, which leaves a fair amount of room for speculation. It is a story concerning the Red Queen, the Red King, and the Court Jester, and it seeks to explain the game of life.

The Red Queen hypothesis proposes that evolution is driven by the competitive interaction of creatures and organisms interacting with each other. As one organism adapts and finds an advantage, others must either also adapt, or failing to, go extinct. The classic model would be a prey species evolving to run faster, such that its predator must match the adaptation or go hungry and ultimately die out. This is a view of life progressing through an 'arms race', and as this competition is constant, the extinction rate and pace of evolutionary change can be expected to proceed evenly over Earth history. The Red Queen hypothesis was first articulated by Leigh Van Valen in 1973, citing Lewis Carroll's character who tells Alice in *Through the Looking Glass* that "it takes all the running you can do, to keep in the same place."

The Red King hypothesis is similarly one of co-evolution, of creatures interacting with each other, however it focuses far greater importance on life forming mutually beneficial relationships. When organisms do so, through various mutualistic interactions, they 'outsource' some of their functioning to another organism. This generates an efficiency, as the organism no longer has to perform all the functions necessary for survival itself, creating benefits elsewhere. For instance, birds in mixed species flocks are able to forage more successfully by outsourcing their predator vigilance to specialists such as Racket-tailed Drongos. As interactions of this kind between organisms form webs of mutual dependency, the complex relationships we find throughout the biological world emerge. From this perspective, evolution can be seen as proceeding primarily from collaboration rather than competition.

Counterpointing these ideas is the Court Jester hypothesis, named after the archetype of the disruptor. In 1972, the palaeontologists Niles Eldredge and Stephen Jay Gould put forward an idea they termed Punctuated Equilibrium, to describe the sudden appearance of new species in the fossil record, followed by often long periods of seemingly little change. This picture of stasis alternating with short flushes of evolutionary innovation contrasts with the steady rate of evolution due to constant levels of competition expected in the Red Queen explanation.

In 1999, Anthony Barnosky proposed a non-biological mechanism for Punctuated Equilibrium, suggesting that bursts of evolutionary development have been prompted primarily by circumstances of the physical environment. These may include climatic or habitat changes due to major volcanic events, continental drift, fluctuations in the sun's luminosity, or the random impact of extra-terrestrial bodies. All these are disruptors to otherwise stable conditions for life on Earth, and hence representative of the Court Jester archetype.

While these theories have prompted much research, a consensus has yet to emerge. Indeed it seems probable that life has developed not by one mechanism alone, but an interaction of them, depending on circumstances.

Going down these rabbit holes into the Wonderland of evolutionary biology can be bewildering, as the details of the living world are so complex. Like Alice, who eventually asked; "Who in the world am I?", I've not only puzzled about nature, but my place in it. For me, answers have emerged from listening. In all honesty, I do not believe I hear expressions of competition. When I do encounter contention, it is most often in the form of a formal agonistic behaviour. Even then, these occasions are relatively rare. The prevalent communications of the natural world I hear as invitations to love and play, bond socially, work together, and to convey individuality, express aesthetics and facilitate essential negotiations.

Reflecting on my years of listening to nature, I have come to a personal conclusion. Most of what we encounter in the natural world can be understood as the outcome of two principles: adaptation and interconnection. Competition, *as we understand it socially*, is not predominantly the way things work for the rest of the living world. Contrary to popular opinion, nature sustains itself through a less contentious agenda.

This may seem a heretical statement – a dismissal of the one foundational idea on which our Western view hinges. And indeed, when I suggest this, I've noticed some people reacting, clinging on, it seems to me, to a competitive paradigm by pointing out specific animal behaviours while ignoring the overall evidence of living systems. I am not surprised by this, it is perhaps too easy to project our own values onto nature. Yet standing back to gain perspective, the strident defence of a belief in competitiveness suggests it as being illusory. Rather than an explanation of the living world, I see 'competition in nature' as a projection of our own assumptions.

Humans have a great capacity to see what we expect to see. If we examine nature through a winners and losers lens, that will inform our understanding. Yet those same observations can be interpreted otherwise. Across the aeons of life, rather than extinction, we can often follow the evolutionary lineages of species morphing into new ones as they adapt to altered circumstances. Even when cataclysmic events end lineages,[57] we can view, as with the birds and reptiles that survived the dinosaurs, nature recalibrating and restoring itself rather than becoming the lesser. In our present day, instead of successful or otherwise, we can appreciate all organisms as being adapted to fitting into the web of life and existing in their own unique ways.

Listening gives us a sensory awareness of what is going on with directness, challenging our assumptions and beliefs. In doing so, we're reminded that, to refashion a sentiment expressed in one way or another by poets and thinkers over the ages, while what we hear is important, more so is the way that we choose to listen.[58]

Interlude

The Immolated Forests

In 2019, the summer fire season began early. A prolonged drought had desiccated eastern Australia, and the bush was tinder dry.

In September, several fires erupted in southeast Queensland, including an unprecedented blaze which destroyed ancient rainforests that hadn't experienced fire for millennia. This was habitat for lyrebirds, logrunners, bowerbirds and scrub-birds. They were also locations in which Sarah and I had done extensive field work to create some of our early albums. ♪ The following day, another blaze began in an area of coastal bushland north of Brisbane. It came within a kilometre of where I had been recording only eight months previously. ♪

This was to be just the beginning. By October, fires had broken out in the Blue Mountains, west of Sydney. Multiple locations I knew intimately were ablaze; Kanangara, Jenolan, and the Grose Valley. A separate fire in the nearby Wollemi Wilderness, where I'd hiked extensively during my youth, was growing into a monster. ♪ By November, massive fires were burning out of control along the Great Dividing Range, consuming vast stretches of forest between Sydney and Brisbane. The Washpool rainforest, where I'd once nearly soaked my equipment trying to record Pouched Tree Frogs in torrential rain, was now so dry that it became part of the conflagration. ♪

Inevitably, we knew people caught up in the unfolding situation. Two of Sarah's friends lived in bushland south of Sydney, and as we checked emergency services information, we realised with dismay that fire had just overrun their locality. That evening, a photo of them both, exhausted and grimy, being hugged by a volunteer firefighter, headlined a news bulletin about their devastated village. At least we knew they were alive.

In late November, lightning strikes began blazes in eastern Victoria. Unable to be contained in remote mountain forests, they continued burning through rugged wilderness areas over the ensuing weeks. By late December, we were despairingly following their progress as they burned their way through the tall forests of East Gippsland. Waratah Flat probably went up shortly after Christmas, in the same firestorm that destroyed Ellery Creek and the Errinundra Plateau a few days later.

As that megafire continued eastward to the coast, the world witnessed dramatic footage of people being evacuated from the coastal town of Mallacoota under apocalyptic ash clouds. Only a dozen kilometres away, a rainforested gully where I'd documented koalas, tree frogs and a rich community of birdlife only a few years previously, was utterly incinerated.

By the time heavy and sustained rains arrived in early February to quench the blazes, and even though the fires never came near where we live in central Victoria, Sarah and I were emotionally exhausted. Feeling fragile and frequently in tears, we comforted each other, unable to fully take in that the places we so treasured were now gone.

All the while, stories of human tragedy were emerging; of lives lost, properties and homes destroyed, businesses wrecked and communities devastated. Meanwhile, Professor Chris Dickman of the University of Sydney had published an initial estimate of the ecological damage – conservatively, a billion animals killed. In time that number would be revised upwards to nearer three billion vertebrates. An unknown number of unique native species were now pushed dramatically closer to extinction; Koalas, Long-footed Potoroos, Platypus, the Kangaroo Island Dunnart, native fish species, and even the Lyrebird. Seventeen million hectares, over twenty percent of Australia's total forests, burned in one season. Many of the cool and wet rainforests affected, which have survived aeons since the time of ancient Gondwana, are not expected to regenerate. The destruction was just too huge to comprehend.

Of course Sarah and I also thought of the Yellow-bellied Gliders, most of whose range had burned, and who could not have quickly moved out of the path of the flames. Similarly those Sooty Owls in their rainforested gullies, sheltering from choking smoke in their tree hollows until it was too late to escape.

We grieved not simply a change in the landscape, but its ruination.

Painful experiences often bring a reconsideration of what is important. I suspect the heartbreak of the bushfires may have reminded many of what is precious – the real values of community and nature.

For myself, it prompted a re-evaluation. The forests of Gippsland, with their cathedrals of trees and echoing with birdsong, were the places that had initially inspired Sarah and I. Their loss, clearly linked to accelerating climate change, was not just an ecological tragedy on an unprecedented scale – it felt personal.

In listening back to the recordings we'd made, I knew that they documented a now-vanished soundscape. This gives them a certain scientific value in monitoring regeneration or as a measure of restoration efforts, however there was little satisfaction in this for us. These circumstances were the opposite of why we'd recorded them.

I also thought of listeners continuing to enjoy our recordings, possibly unaware that the places they represent are so irrevocably changed. Again, this was not the conservation awareness outcome we'd once hoped to convey.

In my own lifetime, I was experiencing an environmental loss of staggering magnitude. Nature seemed so fragile and vulnerable, its vitality so easily lost.

During this time I sought contact with friends. Some had been caught up in events, their properties and livelihoods impacted. Others were ecologists with a more specific knowledge of what had happened. And then there were the ongoing heart-to-hearts –

with old friends, with Sarah. I also went for walks to listen to the birdsong of our bushland, and be reassured.

Everything seemed to circle back to listening. It felt like a bedrock on which to understand, not just recent events, but my own life. I thought of the people dear to me. Each in their own way have shaped me by simply being who they are. However I reflected that they wouldn't have been able to do so if I hadn't been welcoming of their influence.

I've come to consider that the measure of a relationship is how much we're prepared to let someone personally influence us. This constitutes the essence of deep listening; it is not simply hearing someone out, but an openness to take something from the discussion and come away with a broader outlook.

I also thought of Harold, and his instruction to be still and let the bush get to know me, and that if I did, it would talk to me. He was sharing an Indigenous understanding, handed down from generation to generation, a complete and integrated way of both knowing and being in the world. I could now appreciate his words a little better. He wasn't really talking about what I might hear in nature. In a way, he wasn't even talking about sound. He was referring to a humility, a willingness to be deeply influenced by tuning in to the wisdom inherent in nature. For him, listening to nature was a way of learning who we can be.

I've been listening to nature for thirty years now. I've been privileged to travel widely, and hear the universality of the natural world. I've learned something of its wild languages, been puzzled by its ambiguities, and gradually discerned some meaning in them. The bushfires have been a sad reminder that the softly spoken voice of nature is both precious and so, so vulnerable. Faced with a human world absorbed in its strivings, it may also seem inconsequential.

Yet the voice of nature is the voice of the living world. It speaks of the knowledge of the biosphere, with an authority acquired by natural selection acting over immense periods of time. It is expressive of the finely balanced processes that have resulted in the survival and continuance of life. It tells stories of adaptation, diversity, complexity, organisation, relationship, interdependence and sustainability.

Life is an honest teacher. And so the voice of nature is an influence we can trust. Nature is speaking to us of the most crucial matters. It speaks of what is necessary at this time. It is telling us of how to live.

Now, instead of listening to learn about nature, we must listen to learn *from* nature.

Chapter 13

An Ecological Future

We inhabit the world through our senses.
Through our sense of hearing, we commune.
Communication, of one kind or another, allows life to function.
Thus we are each immersed in the living world.

Nature's functioning is the result of evolution shaping life at every level over vast periods of time. Often we think of natural selection as acting on physical characteristics. But it shapes behaviours too. Any behaviour that is not advantageous is discarded by evolution. At the level of the biosphere, natural selection can be seen to have shaped the processes of life as much as individual species. And so, when we encounter expressions of underlying principles throughout nature, we can be confident that it is no accident. Mutual accommodation, agonistic behaviours, the pervasive interconnection of organisms and complex forms of co-operation are all strategies that life has arrived at to maintain itself. They are widespread and foundational in living systems. We recognise them acting now only because they have sustained life successfully over so many aeons. They can be inferred from the deep past, even enduring mass extinction events. They come with an evolutionary stamp of approval.

These 'ways that nature does things' are of crucial significance, as they are the processes by which the living world achieves sustainability. In this final chapter, I'd like to explore how, by listening to nature, we can find insights into our own ways of living.

∗

We are alive at a pivotal moment in our planet's evolutionary journey. For the first time, one species seems to hold the future of the entire biosphere in its hands. Our decisions and actions are no longer of significance only for ourselves, but all of life. It is an unprecedented situation to find ourselves in – a time that demands clear, responsible and visionary action. Instead, confusion, obfuscation, misinformation, polarising ideologies and inaction seem ever more prevalent.

In these circumstances of crisis, I propose that nature has something to say to us. Yet it can only do so if we bring all of our being to listening. We cannot approach this only through objectively analysing what we hear around us. If we rely on intellect solely, it will fail us in this most important enquiry. To listen deeply is to listen passionately, as living entities and co-inhabitants of this planet. It is to resonate with what we hear in nature, finding significance through a sympathetic vibration. It requires that we be prepared to venture beyond what we think we know and consider reasonable.

Perhaps in doing so, we can each respond to the most important questions of our time; what is our human relationship with the natural world, and in honouring that, who can we, and our species, become?

The Costs of Competitiveness

When I was young, my parents bought me a small astronomical telescope. Despite the pervasive light glare of Sydney, I would spend evening after evening in our backyard, with star maps and red cellophane wrapped around a torch, tracking my way from star to star, locating clusters, planets and faint nebulae.

Looking up at the night sky, it occurred to me that the same constellations I was viewing had shone down on our planet for untold millennia. Viewing those impassive patterns of stars, still and unchanging, I had the sense that all our human problems were insubstantial. They are mostly self-induced. We exacerbate our own difficulties. From nature's point of view, our lives don't need to be that problematic.

At the time I was not doing well at school, neither a sporting success nor an academic one, and the realisation gave me reassurance. Looking back, I guess my insight had philosophical parallels, but then it just seemed an 'Oh, I get it' moment.

Life may be far more benign, and living systems more stable, than the nineteenth century 'struggle for survival' assumption has led us to believe. As both the Red King and the Court Jester suggest, the lives of creatures may not be so determined by a daily battle for existence. This morning, I watched as a foraging flock of honeyeaters, thornbills, silvereyes and weebills gleaned through the tree canopy in the autumn sunshine. Witnessing them going about their activity, twittering animatedly, I didn't sense struggle, but that each has its place among a well-ordered web of relationships that sustain them.

Unfortunately, the belief in struggle, and the consequent virtues of competitiveness, have become culturally ingrained and persist through to the modern day. We not only personally internalise them, but they're embodied in the social systems within which we all live. In natural systems, organisms perform specialised roles and frequently enter mutually beneficial relationships. Likewise, we live in societies, performing roles in which we trade our specialised skills, almost always involving collaboration. Why then do we create an unnecessarily harsh system of winners and losers?

When I first arrived in London after my youthful, backpacking travels through Asia and India, I was physically run down and personally knew only a handful of people in the country. I also had few professional qualifications, Britain was in economic recession, and work was hard to come by. Expecting to be relegated to the dole, I went to sign on at a social security office. To my surprise, I was instead offered a job there.

Working on the public counter each day, I processed the documentation of people from all corners of the globe. Mixed in with the local Londoners were Jamaican Rastafarians, Pakistanis,

Irish and Scots, plus Indians like those who'd so recently welcomed me in their country. For some, English was not their native language, others were illiterate, most were confused, disadvantaged and desperate. As best I could, I tried to assist.

Until the day I was called to my manager's office and told with no uncertainty that it was illegal for me to advise a claimant of what payments they may be entitled to. He even dragged out the legislative codes to point out section and paragraph. By law, I was forbidden to assist the people I was employed to help. This was my first collision with the inhumane logic of economic rationalism.

Economic logic is not intended to achieve fairness, provide people with rewarding work, nor support a vibrant society. It is not even a prerequisite of viable commerce. It is simply a competitive logic of numbers, taking little account of people, communities, the preservation of nature or what is required for everyone to live well. It diminishes the central values of being alive.

The chain of events that began with a small astronomical telescope and a childhood fascination with birds and nature, went on to meeting Sarah, establishing Listening Earth and hearing those Spiny-cheeked Honeyeaters at Mutawintji. From this has eventuated a vocation and modest income for us both. We have much to thank our listeners for over the years, and feel privileged to have found work that is fulfilling.

Our vocation has taken us on field trips around the world, not only to document nature, but for both Sarah and I, to immerse ourselves in cultures that are less competitive than our own. My first experience of this was that first trip across Asia and India, 'culture shock' being a curious term for the kind of disorientation I felt. On reflection, my bewilderment was not due to the exotic, which was easy to adapt to. It was in being touched by interactions with people that should have been familiar, but weren't; small kindnesses and unexpected friendliness from strangers, a confidence in the essential goodness of life, and a human closeness that was not native to my own culture.

Sarah and I have since developed an affection for the chaotic vibrancy of India. We've returned several times and made many friends there. It is a warm culture, and I've often found myself taken aback by people's lack of defensiveness. On one occasion, out of genuine puzzlement, an Indian friend quizzed me; "What is it with Westerners that you walk about with walls around you?" I was humbled, not only by his insight, but that he asked the question of me. It implied he didn't see me as being like that, and trusted me enough to give some answer. I hope it was adequate.

My experiences of living among other cultures have shown me that in structurally competitive societies such as in the West, we inevitably become distant from others. We call it independence, autonomy, individuality or privacy, and don't acknowledge that these normalise separation. Sometimes we need space to ourselves. That's healthy. However the psychological defences we unknowingly erect to protect ourselves from rivalry – self-importance, cynicism, ego, suspicion, hubris, a loss of faith and trust – are not.

The costs to the psyche of diminishing others through social competition are unavoidable. When our natural openness, trust, mutual care for each other and a sunny faith in life are lessened, we lose something of our soul. The ultimate price we pay in a competitive society is our natural birthright to mental wellbeing.

Many would assert that it is in our nature to be competitive. We enjoy it, you can't stop us from wanting to win. Sometimes it is for the sport, at others it becomes serious and all absorbing. Games are how life is played.

Maybe, but like competition in nature, I conclude the belief that individuals are naturally competitive is overblown. In reality, humans are among the most consciously co-operative and altruistic species on the planet. Working together for a common good seems the superpower which evolution has endowed us with. In moments of crisis, we can be extraordinarily supportive of others in need, even to the extent of risking our own lives.

This natural sociality has manifested an abundance of societies that are far less competitive than our own. Many indigenous and traditional cultures, like those of Australia's First Peoples, have demonstrated their viability for extraordinarily long periods of time. Common to many of these societies are rituals that can be invoked to resolve conflict, formal practices that represent our own species' agonistic behaviours.

And there is another strong repudiation of the belief that competitiveness is inherent in human nature – we have to teach it to our children. Throughout an intense educational process lasting many years, young people are prepared for life by constantly ranking their performance. It is not only on the sports field that they learn the value of winning. Continual academic testing, culminating in stressful examinations, pressures them to achieve.

The years of childhood are the most formative and influential of our lives. It is when young minds are wide open, soaking up the world, forming interests and learning about friendship. From my own youthful experiences, I know that enforced competitiveness corrupts this natural learning process, replacing the open fascination and curiosity of children with an obligation to perform. It is also when children are the most vulnerable to the rewards or disapproval of those they trust and love. To lose is to disappoint. To fail repeatedly is to be sidelined. It's easy to see how a potent emotional mix of love, acceptance and inclusion can grow into an adult addiction to winning.

All this preoccupation with teaching competitiveness is justified by one word: success. We want our children to be successful in life. Assured, independent, skilled, resourceful, responsible, confident – these are good personal qualities, and I wish I'd emerged from my schooling feeling a little more of them. Yet these strengths of character can be far better fostered through affirmation than contention. And when you reduce the justification of personal achievement to its core, it is our competitive society that young people are being educated to be successful in.

So here's the thing – if competitiveness is not 'natural' but cultural and learned, then it can be unlearned. We know the human mind is extraordinarily plastic. We can grow new neural pathways. We can adapt and develop our capacities throughout life, and reimagine ourselves. Instead of being limited to a competitive approach, we can educate ourselves in a much broader range of life skills.

If we are to develop toward a less competitive society – and to address our current crises I believe we must – then this is surely the most hopeful way we can nurture the generosity of the human spirit. The best gift we can offer young people is to discard the pressure cooker of competitive education, and instead encourage them to follow their curiosity, develop their talents, acquire skills, explore their interests, and allow them to grow up naturally.

Restoring Agonistic Practices

Such an optimistic view may suggest that I'm coming down on the 'nurture' verses 'nature' side of a much-argued debate. This controversy was given impetus in the late 1970s by disturbing observations made by Jane Goodall's team of primate researchers at Gombe in Tanzania. Wild chimpanzees were witnessed to be capable of wilful viciousness toward their own kind, co-ordinating lethal attacks on neighbouring groups, brutally attacking their own kin and even cannibalising infants.[59] The conclusion that human competition and aggression appears related to this primate behaviour has been profoundly sobering.

However *related* doesn't necessarily imply *inherited* as a behaviour. Instead of being due to 'malevolent genes', the violence of both chimpanzees and humans can be seen as quite the opposite. Rather than genetically blueprinted, it may be an inevitable consequence of intelligence.

Complex intelligence is rare in nature and, depending on how you measure it, likely unique to our primate-homo lineage. Intelligence gives us the ability to conceive of things that do not exist in reality, and to manifest previously unimagined possibilities.

More significantly, intelligence allows us to transcend biologically inherited patterns of behaviour.

Among these inheritances, shaped by the evolution of generations of animal life before us, are the agonistic behaviours which minimise aggression. For most animals, these are genetically encoded, specific to the species, and invariable. Intelligence however has given us the agency to over-ride our natural agonistic behaviours, allowing us to conceive of harming our fellow humans if we choose.

This places human aggression on the same side of the 'nurture/nature' argument as competitive behaviours, and suggests a direct link between the two. There is a body of evidence to support this view. Social scientists have put forward the idea of 'cultural speciation', or pseudospeciation, to describe the way that humans have a tendency to form identity groups along shared ethnic, national, ideological, socio-economic or religious lines.

This is all too familiar. As Sarah and I travelled around rural Turkey, we'd ask about the next town we planned to visit, often finding locals shaking their heads in distrust of their neighbouring townsfolk and advising us to be careful. Upon arriving, we'd find the same said of where we'd just been. This may seem a trivial example, but as the process of cultural speciation intensifies, distrust can grow to an exclusion of those defined as the out-group, ultimately paving the way for violence. Not far from those quiet towns we visited, the dehumanisation of others has led to the Armenian genocide, the ethnic cleansing of the Balkan wars, and the ongoing Syrian and Kurdish conflicts.

In 1986, an international meeting of scientists was convened in Spain to adopt a statement to "refute the notion that organised human violence is biologically determined". Specifically, they asserted that it is scientifically incorrect to say that we inherit violence from our animal ancestors, that it is genetically encoded into human nature, or that evolution has selected aggressive tendencies in preference to any others. Their conclusion is that violence, far from being natural, must instead be incited. The Seville Statement on Violence was subsequently adopted by

UNESCO, and has since been widely disseminated and endorsed by major scientific bodies.[60]

Listening to nature provides a further perspective from which to consider human competitiveness. Natural systems are governed by complex interactions, many of them generating feedback loops that regulate toward balance and stability. We've encountered these in the afternoon hush of the Papuan cloudforests.

By contrast, our human systems are largely founded in amplifying feedback processes. Businesses are obliged to seek profit without restraint and economies are geared to grow, while personal desires for wealth accumulation rarely seem to be self-limiting. In governance, a lust for power and control can equally become all-consuming. At the other extreme, poverty can intensify into an inescapable spiral of destitution and despair. We do have disruptor mechanisms to break these amplifying processes – legal, social, charitable and governmental interventions – however they are often invoked only to address situations that have already reached a crisis.

In nature, regulating feedback processes have evolved within living systems. Humans have developed our own life systems, among them politics, economics, law and culture. We can view these as true analogues of natural systems, as they exist to achieve the same ends: the maintenance of life within a complex community. Yet our social values endorse, and our institutions are set up along, principles entirely the opposite of what sustains nature. While nature ensures stability through self-regulating processes, we continue to celebrate extremes of wealth, status and power, despite knowing full well the consequences.

How would our lives be if we could avoid the pitfalls of systemic competitiveness and runaway inequity in the first place? More to the point, how could this be achieved?

Traditionally, we've relied on moral systems to moderate social behaviour. However history shows us that moral codes alone are insufficient to prevent violence, and religious and ethnic identities frequently play enabling roles in defining 'us and them'. Likewise, we also know that strong leadership can be so easily corrupted, and charismatic rhetoric employed to incite division.

If good leadership and compassionate ethics are to fulfil their potential to benefit us all, they must be enacted through social institutions that function akin to the negative feedback processes in nature – limiting contention and the failings of human ambition. Given that we live in pluralistic societies, possibly never more so than now, only secular institutions offer this hope, as they are ones in which we can *all participate as equals*.

As with nature, civil society is not a noun, but a verb – it only works when we engage and contribute to it. Through doing so, we can shape governance to less gladiatorial agendas, and our economies to achieving sustainability, enabling us to adapt and exist comfortably within the resource limits of our planet.[61]

Civil society, by its founding principles, also promotes cultural diversity. My mother used to acknowledge this with the observation that "it would be an awfully boring world if we were all the same". Like biodiversity, human diversity is to be encouraged and celebrated. So rather than allowing cultural speciation to drive wedges between us, we can promote the equity and inclusion required for us all to live together safely.

Importantly, both ecosystem feedback processes and agonistic behaviours are often facilitated through communication. If we are to learn any wisdom from nature, it must be to similarly communicate, and to strengthen our civil institutions such that they enable the agonistic rituals of our own species: diplomacy, negotiation, inclusion, truth telling, reconciliation, fairness, tolerance, trust and kindness.

Listening to nature has helped me appreciate that civil society is very much based on similar principles to those that sustain nature – a parallel which not only highlights the importance of progressive social ideas, but tells us *why* they work.

Considering the state of the world currently, manifesting these ideas may seem a hard ask. Could humankind have such a change of heart? I'm hopeful. We've changed so much already over the course of our history, and there's little reason to believe we won't continue to. And events are moving quickly now. Indeed, I see many signs that a shift in sentiment and yearning for a better world are already under way.

Ultimately, my positivity is founded in our inescapable relationship with nature. A study of history reveals that the environment has always shaped human civilisations, and continues to do so, whether we acknowledge this or not.[62] In due course (and if nature allows us to survive long enough), I suspect the reality of our existence as living creatures will oblige us to adopt humane and ecologically respectful ways of governing, trading and structuring our societies – ones that, through a process of biomimicry, learn from and are modelled on nature's systems.

Our Ecological Purpose

The human evolutionary journey began, like songbirds, with innovations in how to communicate.[63] From primate origins, we have become social and learned how to work together. We've now pursued this path to become as close to a dominating species as our planet has ever known. Yet in doing so, we are destabilising the biosphere and bringing it to the potential of collapse.

Understandably this realisation has precipitated alarm, yet humankind's disconnection from nature is a hard habit to overcome. It is this pervasive failure to recognise ourselves as integral to the living world that many point to as being at the heart of our current environmental crisis.

One seemingly trivial manifestation of this disconnect is expressed in popular culture, in the idea that 'nature would be better off without us'. This particularly peeves me, the idea that somehow, if we were to annihilate ourselves, nature would simply return to its previous abundance. Dystopian stories of our future have become so ubiquitous that, as has been pointed out, it is easier for us to imagine the end of the world than envisage a future in which we embrace positive ways of living.[64] Perhaps it is a contemporary expression of the apocalyptic, 'end times' beliefs of Western traditions, but to me it just seems defeatist, and to deny that humankind actually has an important purpose to fulfil on this planet.

Many 'nature after us' scenarios also have a serious flaw in reality. We've destabilised the climate already, stripped fertility from soils, introduced long-lasting toxic pollutants into the environment, compromised ecosystems and decimated wildlife populations. A meltdown of the nearly six hundred nuclear reactors currently being maintained in operation around the world would be a final, ruinous legacy. A catastrophic collapse of our technological civilisation would equally condemn the biosphere. The Earth cannot afford *Homo sapiens* to be a failed evolutionary experiment.

So humankind has a job to do in restoring the living world and maintaining the Earth's systems – a task that only our species can undertake.[65]

My conception of this is that we need to identify and embrace an ecological purpose for our species. Every one of the biosphere's multitude of organisms perform roles within ecosystems, contributing to preserving the balance of life. Now is the time we must find ours. After a long history of traditions that view humankind solely in terms of moral actions, we are coming to understand ourselves as ecological actors. Our habitat has become the entire planet. In spreading out to occupy every corner of the globe, we've come to believe we own the joint. We take without thinking of giving back. This is not edifying. We can do better. And we also must, as we now know the degrees to which we're compromising the functioning of the living world on which we depend.

In finding an ecological purpose for our species, I'd concur with the many thinkers who have suggested that it will be characterised by a spirit of stewardship. Here, listening can guide us. It can begin by reminding us of an essential truth: every organism is a sentient being, each in its own way. Some are soniferous, and we can hear them telling us of their aliveness. Others are non-soniferous, but nevertheless communicate, are aware of their surroundings, and have their own expressions of sentience. Listening calls on us to embrace all living things as having a right to survive.

Beyond this obvious truth, listening can provide valuable information. For instance, in restoring natural landscapes, acoustic monitoring for the presence of species can complement listening for more subtle clues to ecosystem integrity. The density and diversity of insect songs will point to the state of the food web's foundations. The coherence of dawn choruses will tell us whether birds are interacting as well-sustaining communities. The dialects of honeyeaters may allow us to map their movements and pollen dissemination across arid landscapes. Temporal cycles of sound and stillness can provide an overarching measure of well-balanced and functioning ecosystems.

Sadly, in the mere handful of decades I've been documenting the natural soundworld, this delicate tapestry of interactions and negotiations is noticeably unravelling. I don't readily find the nuanced soundscapes I did when I first began. Many of the sonic behaviours I describe in this book are less easy to encounter now, replaced by creatures trying to adapt to altered circumstances.

If we're to turn this around, the restoration of wild soundscapes can guide our efforts. Nature tells us stories. If we know what to listen for, we can both hear when living systems are well, and when they're not. Like the tuner of a piano, we can listen with an informed ear to discern exactly what a vibrant ecology should sound like. We can also become aware of the technological noise we discard into the world, and by moderating it, lessen the acoustic stress on both wild creatures and ourselves.

This is the ultimate value of listening: If we get the sound right, much else will fall into place.

However an analytical approach to gathering acoustic data about the natural environment, akin to a doctor diagnosing a patient, has its limitations. The information we glean may tell us much, but what do we do with this knowledge? Western beliefs are founded in disconnection and assume an entitlement to manage, extract and exploit nature. We learn these human-centric ideas through habituation and socialisation. If we're to act differently, we need a new perspective, one which can *only* originate from outside the human world, through opening ourselves to nature.

I'm not the first to say this, but at its heart, the environmental crisis is actually an identity crisis. It results from our collective failure to understand who we are, and our relationship with the natural world. And so we need an identity refresher. Without re-imagining our species' purpose on this planet, we will continue, in one way or another, and possibly despite wishing otherwise, to perpetuate the same outcomes.

With this understanding, listening to nature becomes not simply the hearing of sounds, but a means of renewing our place in the world. For modern peoples, deep listening can be what it has been for our ancestors, and continues to be for Indigenous peoples; a communion between the human and natural realms.

Chapter 14

Hearing Our Place

Wherever Sarah and I have travelled to do our field work, we've always found solace in wild places. Even when cautious about wildlife, we've nevertheless felt comfortable in natural surroundings. This ease of feeling is what I suspect Aboriginal peoples wait for when being welcomed by the land itself, as our guide did at Murujuga.

Which makes it all the more disquieting when one doesn't sense it. This happens rarely. On one such occasion in southeast Australia, tired after a day of driving, we left the highway and followed a dirt road into forested hinterland in search of a bush camp for the night. In the headlights, we looked for somewhere to pull off, eventually finding a rough, narrow vehicle track diverging between the trees. As we set up our tent in the dark, we both felt the place to be eerie, unsettling, but couldn't put our finger on why. Exhausted, we slept.

The next morning, I awoke at first light and walked further along the track. It ended only a few tens of metres on, opening onto a vast, bare area of ground that stretched into the distance – a clearfelled logging coupe. The tracks of heavy machinery scarred the muddy earth, tree debris littered the ground and shattered stumps stood as memorials to the ecosystem that had once been. No birds sang. Even the paling sky looked sombre. No wonder we'd felt uneasy; the land was devastated, there was no living thing to welcome us.

That humans are capable of such an instinctive knowing of life around us speaks of an essential connection we have with nature. Wildness has a presence to it, and we sense it. In natural surroundings, I often find myself dropping into a calm and curious awareness. It seems to me this state of being is indistinguishable from that cultivated in mindfulness meditation, where the focus is on becoming aware of one's inner life. When listening, I feel an inexplicable connection between stillness, awareness, and the wellbeing of both ourselves and the natural world. Earlier I spoke of this as hearing the mind of nature. Of all our senses, listening brings our attention to the inner life of nature.

It is not surprising then that many find quiet time spent in nature to be restorative. However it is not only ourselves that benefit. Our communion with the wild will rub off on others too. This is an important realisation, particularly for those of us who worry that our individual contribution to changing the world is insubstantial. Instead, we can focus on allowing the natural world to change *us*. Returning from our sojourns, we can be quietly confident that what we've absorbed from nature will permeate others. It is an inevitable influence. It cannot be otherwise.

You may perceive your connection to nature to be personal, but it doesn't have to be experienced privately. The next time you're in a natural place with friends or family, you may like to suggest sharing a listening walk. This is a simple yet powerful thing to do, bringing you into the moment and awareness of your surroundings.

Before you begin, set an ending location, somewhere you'll walk to, but not far, as you'll be proceeding very gradually. Then, without speaking, placing each foot mindfully after another, one deliberate step at a time, walk very slowly together, feet aware of the earth, ears attentive to everything around you. Try some of those listening awareness exercises I spoke of earlier. Experience that quietness that connects you with the presence of nature. Let yourself mellow, opening, abiding together for twenty minutes, maybe more, allowing some time to sit quietly or share at the end. We're usually so active and communicative, babbling away to one another, so this is a very special way of bonding and connecting. And if you're with children, let their imagination remind you of how to listen playfully.

These examples of personal and shared listening can be expanded into meaningful ritual. In Australia, the Aboriginal custom of welcoming guests is known as a Welcome to Country. This ritual is now frequently offered by Aboriginal people at public events to those of non-Indigenous background such as myself. When hosting, a non-Aboriginal person may give an Acknowledgement of Country, stating the Indigenous name of the place and offering respect to the elders and traditions of this land's first peoples.

In addition to these, there is another practice that I propose we may consider adopting. Ultimately, all our human interactions exist within our relationship to the natural world. It is nature itself that has to welcome us, as our guide at Murujuga practiced, and Harold suggested to Sarah and I all those years ago. While this understanding survives in Indigenous traditions, Western peoples have long ago abandoned it.

So as well as sharing a Welcome *to* Country, or an Acknowledgement *of* Country, we can open ourselves to a Welcome *by* Country. When you go out into the bush, alone or with others, as you first arrive, take a little while to pause quietly. Wait to be welcomed, and develop an instinctive sense of recognising how nature may signal this to you. It could be a bird flitting in to check you out, a moth hesitating momentarily while fluttering nearby, or a patch of native orchids nodding in an imperceptible breeze. There is no formula here, you will know when you feel it, and can proceed with your activity. This simple and perhaps personal ritual goes to the heart of our modern lives, allowing us to find ourselves anew in the presence of nature.

However you approach it, there are no right or wrong ways of listening. There are no correct interpretations of what one hears. Deep listening to nature is radical, even subversive. It takes us out of our human bubble, and allows us to directly hear the voices of other beings. If you are sensitive, if you listen with heart and mind, their voices will speak to you. Please welcome this. Be open to nature revealing itself, and telling you things. As you tune in more deeply, I hope you will find, as I have, an accord between the ways of nature and your own intrinsic human nature.

Ultimately, I find this the most affirming realisation – we do indeed belong as part of the Earth family.[66] Our senses give us an intimate awareness and knowing of the living world. Connection with nature is one of life's good things. So be still and listen. Take your time. Extend your senses. Let nature get to know you, and in its own way, to welcome you.

Coda

It is a sunlit autumn morning as I take a walk along our bush track. A gaggle of Red Wattlebirds are moving through the tree canopies, their harsh calls like perky detonations going off here and there. A White-bellied Cuckoo-shrike flies high overhead, its thin cries echoing off the sky. I notice a soft hum emanating from the crown of a eucalypt tree and see it is in full flower, adorned with clusters of pale yellow, feathery blossoms. Native bees have discovered its bounty and are studiously working the canopy.

In recent weeks I've encountered foraging flocks reconvening, and further on find myself surrounded by a little bustle of activity. There are Buff-rumped and Brown Thornbills exploring a low shrub, a few Weebills gleaning in the lower canopy, Sittellas probing at bark, Spotted Pardalotes higher in the crown, and a Grey Fantail, which pirouettes briefly in the air in front of me before flitting off. Some of these tiny birds are too quick and plainly plumaged for me to identify by eye, so I'm recognising them by ear. The pardalotes are particularly vocal this morning, their contact calls being sweet chimes which emanate softly from nearby treetops.

A little further on, I disturb a family of choughs on the ground, which squawk and flap up into the canopy at my approach. Their whistled 'yellow alarm' signals are pitched high in urgency. I respond, mimicking, whistling back to them but intentionally pitching my notes lower. "It's OK, no need to fuss." This seems to calm them, and they soon glide back down to the ground, their final whistles matching my relaxed pitch.

As it is autumn, I'm anticipating that any day now I may overhear our Scarlet Robins giving their late season song variation. I wonder if I'll hear it again this year, and whether it will shed further insight into an audible link between them and their Red-capped Robin relations? Last summer, a red-cap turned up in our woodlands for a few days, as they occasionally do during the hottest months. On two consecutive mornings I observed it perched in the same tree as a male scarlet. The two were singing, each giving their own unique songs, apparently accepting the presence of the other. Even among such familiar species, their lives continue to be mysterious to me. 🕊

Distracted by my thoughts, I notice sweet fragments of melody drifting from an adjacent gully. I pause, momentarily puzzled. I know every song of these woodlands, but it takes a few more phrases for me to identify it as a Speckled Warbler. They are resident here, but in such low numbers that I infrequently hear or see one. Excited, I turn off the path and walk toward the sound. I don't have my usual sound recording gear with me, just my phone, but it'll do. It's only mono, and the microphone is not sensitive to the complex sonics of birdsong, but I can at least document the moment. I set it running and continue wading through the waist-high cassinia bushes.

I approach carefully. The warbler is still singing; such tuneful whistles, a little poem of song being given every so often. It is actually quieter, and the bird closer, than I'd thought. I stop and wait, noticing a cassinia twitching with hidden movement. Now the warbler has switched to harsh buzzing calls, curious at my presence. It appears at the top of the cassinia; a feisty ball of feathers, all delicate speckled markings regarding me with a stern eye. I feel privileged to view this rarely seen resident so closely.

It dives down into the foliage and begins circling around, still rasping noisily, to reappear by my side for another look. A short flit and it's up onto a low branch, where a family of Superb Fairy-wrens have also appeared, chipping quietly among themselves. A Yellow-faced Honeyeater now sweeps in too, and I feel quite the centre of attention, standing motionless with my phone held out toward them. Curiosity sated, they each move off, the warbler dropping into the bushes to presently resume its song. 🕊

Leaving them to continue their day, I return to the track. At the top of the ridge, by the side of the path, grows an ancient eucalypt. It is a Red Box which we estimate is possibly 400 years old. It's a gnarled specimen, robust yet twisted, its trunk covered in burls and crown blown out by some forgotten windstorm. I regard this as an elder tree, and occasionally stop to rest my hand on her rough bark.

I've been told that in some Aboriginal languages, the words for bark and skin are the same. I think of the wisdom of this land, and my efforts to tune into it. A Tasmanian Aboriginal man once said to me "You can't expect the bush to talk to you unless you talk to it too." So I often chat with this old tree, even if just to wish her well in passing.

Looking up now, I see, beyond her greenery, the blue hemisphere of the sky. Today the weather seems benign, and climate change theoretically remote. Yet these woodlands bear the scars of successive droughts and insect imbalances, with some trees having lost foliage in the recent past due to these stresses. I need to be prepared that, if climate change accelerates, drying and the heightened possibility of wildfire may change this place as irredeemably as Gippsland's forests.

Beyond our precious atmosphere, invisible by daylight, lie the unchanging stars that fascinated me as a child. Those stars are often presented by science fiction as our future, but I believe this to be little more than fantasy. If we are to survive, our future will have to be more ecological, less technological. It will not be the projection of our colonial past into space races, interstellar wars and galactic empires. Rather than mixing with aliens in cosmic bars, we will come to find greater pleasure in celebrating our human diversity, and marvel at the alienness of organisms with whom we share our own planet. Instead of terraforming other worlds, we will find greater purpose in caring for the Earth's life support systems. Forgoing a seat on the flight deck of the Enterprise, we will humbly take our place as just one species among many, all unique, and all with a vital role to play among the wonders of life here on Earth, the only place we truly belong.

Finishing my chat with the old tree, my hand slides away from her bark. In the same way that humankind must, I turn to walk the path homeward.

Acknowledgements

At one point in the writing of this book, I ran out of puff. I found myself wondering why I was pursuing the project at all, and whether it had value.

Then, during one of my morning walks in the bush, I became aware that the ideas I had been writing had not come from me. They'd most often occurred when I was outdoors, contemplating something I'd heard, wondering how life has brought us all to this point and what was to be learned. I realised that 'my' ideas were not being conceived, but received. I didn't know where they were coming from, but they seemed associated with being in the bush and immersing myself in listening. Without this, they wouldn't materialise. Hence, if the ideas were formed in nature, then my role was as the messenger, translating and conveying them for you, the reader. This was a huge relief – I was no longer writing for myself. Instead, I was writing to honour what I had heard over many years in the bush. And so, for want of a better way to say it, this book is my thank you to nature.

Of course there are many people I'm honoured to acknowledge also.

The journeys in these pages have not been mine alone – Sarah has shared almost every step of them. While I've been sound recording, she has been behind the viewfinder of her camera, making carefully composed and beautiful images. Earth to my air, she has also organised travel, run the practical side of our business and been a wise judge of character at times when we've had to rely on others.

Together, we've shared moments of transcendent joy and great danger. On one occasion in Africa, being driven by Roger, we were exploring slowly along the banks of the Ruaha River. Unexpectedly, a huge male elephant emerged from a thicket of vegetation and began charging us. In India, we'd experienced mock charges from elephants, but with ears out, tusks down and barrelling toward us, this animal was not kidding. It was only Roger's quick reflexes and driving skills that saved us. Safely back at camp, he was shaken and uncharacteristically sombre, confirming how close we'd come to tragedy.

During that trip, I also recall Sarah's delight on first encountering wild giraffes, a mother with a calf resting at her feet. On our final evening, we watched together by moonlight as a trio of elephants moved like phantoms through our bush camp, making their way to feast at a favourite fruit tree.

Incidentally, I must tell you of new directions that have emerged for Sarah in recent years. After our decades of field work together, she has embraced a childhood dream: to become a potter. Studying with a master potter locally, she has now set up her own small studio and gallery. With a certain amount of partner pride, I can say that she is very talented at her new vocation.[67]

So in many ways, this book is as much hers as mine.

When we were at Mutawintji, it had not occurred to me that I may have colleagues. It was only after we returned that I met other nature recordists. Amidst cassette and reel-to-reel machines in the home studio of Rex Buckingham, I became aware that others in this country had been field recording, some over many decades. When Rex identified a Red-backed Kingfisher clearly audible on one of my recordings – a beautiful bird I'd never set eyes on, and yet had unknowingly been in the presence of – I also realised how much I had to learn.

Through him, I became involved with the Australian Wildlife Sound Recording Group. On first attending one of their biennial conferences, I met individuals of that pioneering generation who did groundbreaking field work, often with the most primitive equipment and under arduous conditions. I recall one of them, Harold Crouch, eagerly taking me aside to share his fascination with the dawnsongs of White-plumed Honeyeaters. His curiosity fed my own, and I wish he were still around to discuss the significance of pre-dawn repertoires.

Also during that gathering, a session was convened to formally incorporate, which involved deciding on a name. We were Australian wildlife sound recordists, but were we a club, an association, a society? After much discussion, another of the old hands, John Hutchinson, softly suggested that, when it came down to it, we were a group of friends. And so the AWSRG we became.[68]

John was right; in addition to colleagues, many in the AWSRG have become close friends. I've learned from them, travelled with them, and have so much to thank them for. I can't name them all, but they know who they are. In turn, they have now elected me their president, a questionable honour as it indicates I'm now becoming an elder of the tribe myself.

Meanwhile, my overseas travels have resulted in connection with colleagues around the globe. In particular, I wish to acknowledge Geoff Sample[69] and Lang Elliott,[70] both of whom have been publishing stereo soundscapes for as long as I have. Geoff is not only a sensitive recordist, but in his thinking, blends a long history of European cultural knowledge of nature with contemporary biological science. Whenever we've been in the UK, he and his wife Jane have been generous hosts, and I've enjoyed hours in his study, comparing experiences and recordings from opposite sides of the globe.

Similarly, Lang has swept us up during our trips to the US, taking us hiking, canoeing and exploring his local Adirondack Mountains in upstate New York. Our first meeting was enlivened with his enthusiasm for the SASS microphone system, a technology originally intended for recording classical music, but one he found rendered a rich binaural portrait of the landscape. Lang subsequently arranged for a unit to be customised for me, and it has now travelled the world, capturing the majority of my album recordings. I still use it to this day.

For equally collegial friendship and hospitality, my heartfelt gratitude goes to Doug Quin, Peggy Droz, Gina Farr, Jim Cummings, Dan Duggan and Sharon Perry in the states, Mark Brennan in Canada, Marc and Olivier Namblard in France, Roger Boughton in the UK and Juan Pablo Cullaso from Uruguay.

I also want to send a shout out to our wildlifer friends in India, in particular Kishore Gumaste, Rahul Rao, Girish Vaze and Kedar Bhat. Through his nature travel company, Foliage Outdoors, Rahul and his team waded through baroque Indian booking systems to arranged much of our travel around the subcontinent over several trips. He also organised, at very short notice, several public talks for me, even conveying me to one venue on the back of his motorcycle through Pune's peak hour traffic – not a very calming start to the event.

Our guides and drivers in various countries deserve far more thanks than I can offer here. From the stories of our time with Roger (Rajabu) Kissaka and Ally Kimea in Tanzania, and Twomey and Moffat in the Solomon Islands, you get a sense of how indebted we are to them. In India, Thailand, Vanuatu, Nepal and elsewhere, our local companions have done far more than simply convey and guide us around to make sound recordings. They've allowed us to glimpse the hardships of living in poorer countries, and often welcomed us almost as family. To Shankar, Jyoti and KB, Shiva, Suchat, Pak Martin, Sanjeet, and others whose names are now lost to us, I offer our salutations and deep appreciation.

As an example of the ways in which we've touched each other's lives, I'll mention Dharmaja, our driver at Nagarahole National Park during our first field trip to India in 2002. As the local village jeep driver, he had never entered the nearby park, and yet he not only rearranged his daily schedule to facilitate our early morning safaris, but turned out to be an adept wildlife spotter. We spent ten days with him, and on our final morning he brought his wife to join us for her first experience of the park. On another trip many years later, in the Foliage Outdoors office, Rahul mentioned that they had found a great driver for their Nagarahole tours. We were delighted to learn that it was none other than Dharmaja. Ours was an unlikely connection – he was our first driver on our first overseas field trip, and we inadvertently introduced him to his future work.

Our drivers have not been the only people to go out of their way to assist us so far from home. In Bangalore, we were introduced to a prominent wildlife conservationist, known to his friends simply as KN. After an evening talk at his home, and the following day spent with his family, he took us to visit his rural estate, which he ran as a sanctuary. From there he made arrangements through his contacts for us to be accommodated at another park. This spontaneous helpfulness reflected the dignity and inclusiveness with which he seemed to approach everyone. Weeks later, as we were leaving the country, I phoned him from the airport to thank him again and say farewell. I'm glad I did. Only a fortnight later we received word that he had died in an accident on the land he loved.

This book has grown out of more than international travels and ambles through my local bushland. The narrative of ideas has emerged gradually over the years, largely through invitations to give talks and lectures. Whether they've been to small community groups or as a conference keynote speaker, I've taken the opportunity to continually hone ideas and refine my presentations. In particular, I'm grateful to several academics who have extended the opportunity to work with their students, including Leah Barclay, an acoustic ecologist formerly at Griffith University, plus Gregg Muller, Alistair Stewart and Philippa Morse of La Trobe University, Bendigo, who have invited me to guest lecture each year to their outdoor education class.

Along the way, there have been other significant engagements, including a series of presentations at Woodford Folk Festival, which were recorded by ABC Radio National for their *Big Ideas* program, plus a TEDx talk in Canberra.[71] The latter forced me to be concise and selective about what I considered important, an attitude I hope I've transferred to this book.

In 2019, I was invited by Dr Michelle Maloney to speak at an Australian Earth Laws Conference.[72] For the first time, I realised I would not be addressing a nature audience, but one comprising lawyers, ethicists, social academics, activists, policy administrators and the like. Instead of delivering my conclusions about what we could learn from nature with a wink and a raised eyebrow, I was going to have to spell it out. Even though the concepts I presented on that occasion were a rough sketch, the audience's keen interest demonstrated that I was addressing a topic of value, and I've continued refining my thinking. The final chapters of this book are the result.

Once the early manuscript was complete, and with some trepidation, I sent it to colleagues and friends for their feedback. I am particularly indebted to Jennifer Ackerman for her encouragement when I was doubting myself, plus her author's advice and industry advocacy. Sally Polmear and Brian Walters took the time to forensically examine my text, each offering a great swag of the most constructive suggestions and comments.

From her perspective as both an ecologist and sound recordist, Dr Sue Gould has been of great assistance in identifying areas where further enquiry was required. Geoff Sample's feedback, particularly regarding his native Europe, I've taken with great consideration. Gregg Muller and Alistair Stewart have contributed to the section on Grey Currawongs and Cherry Ballarts. The author and naturalist Tim Low has contributed to my thinking, particularly in chapter five, and picked up issues that required clarification. Peter Yates and Craig San Roque's understandings, gained through extensive work as anthropologists among the Aboriginal peoples of central Australia, have been helpful, and John Smith Gumbula, a Wakka Wakka man and advocate for Indigenous arts and culture, has voiced approval of the manuscript.

Richard Sullivan, Dr Lynne Kelly, Craig Morley, Jan Wositzky, Richard Weis and Ted and Jenny Kent have also given feedback and support that I've found invaluable beyond measure. Finally, Julie and David Gittus have offered structural feedback that this book has greatly benefitted from.

Once satisfied, I forwarded the manuscript to Kristin Gill of Northern Books, a freelance literary consultant and event organiser, for her assessment. Her positive encouragement and practical assistance gave me the lift to take this project toward completion, and I'm not sure how I would have done it without her. Robin Murdoch edited the final manuscript with an eagle eye, and Helen Christie refined the design, typeset the book and prepared it for printing.

The cover features an evocative photograph by Lachlan Read. Whilst it is relatively straightforward to take portraits of birds and animals, to capture wildlife images with mystery and beauty as Lachlan has done, is a rare art indeed.

My heartfelt thanks to all these folk who have assisted, contributed, challenged and supported me. If I've omitted anyone in these acknowledgements, this is more my oversight than any lack of appreciation.

Finally, my gratitude to you, my reader, for taking the time to share in this listening journey. I hope it inspires your own, and welcome any conversation that may emerge one day.

Notes and References

Online materials are indicated by ⚭. Links are given on the same website as audio for this book: <https://listeningearth.com/deeplistening/>

Prologue

1. Whilst spelt Sarah, it's pronounced Sara, in the European way.

Chapter 1: An Invitation to Listening

2. Pauline Oliveros. (2005). *Deep Listening: A Composer's Sound Practice.* iUniverse. ⚭
3. Hildegard Westerkamp, Inside the Soundscape. ⚭
4. I would like to acknowledge that not all of us have perfect hearing. If this is the case for you, I nevertheless hope that you can read past the assumptions necessary for this book, and find value in what is written. In my understanding, hearing aid technologies can assist in the field. One of my friends, after decades of hearing loss, has found that good aids have made small birdsong audible again. Another, who is not only a fine naturalist but a professional nature sound recordist, has had significantly limited hearing all his life. He has collaborated in developing a device to reduce frequencies to an audible range, however this unit is unfortunately no longer available. ⚭
5. Lang Elliott, Music of Nature. ⚭
6. Lynne Kelly. (2016). *The Memory Code,* Allen and Unwin, ISBN: 9781760291327.
7. R. Murray Schafer. (1993). *The Soundscape, Our Sonic Environment and the Tuning of the World.* Destiny Books, ISBN: 9780892814558.

Chapter 2: A Practice of Listening

8. For Australian frogs: FrogID app, from The Australian Museum. ⚭
9. Australian cicadas: Lindsay Popple's cicada site, ⚭ cicadamania. ⚭
10. Global birds audio: xeno-canto. ⚭
11. Global birds audio: Cornell Lab of Ornithology eBird. ⚭

12. Australian birds audio: the 'Morcombe and Stewart Guide to Birds of Australia' includes audio from most species, as does the 'Pizzey and Knight, Birds of Australia, digital edition (PK Birds)'. 'Stewart Australian Bird Calls' offers more examples covering subspecies and dialects. All are available via Apple and Android stores.

13. Bird field guides with good voice descriptions: this is a very personal assessment ... For Australia, *Graham Pizzey and Frank Knight's Field Guide* is accurate and evocative (sometimes even amusing). The more recent *CSIRO Australia Bird Guide* by Peter Menkhorst, Danny Rogers and Rohan Clarke is also clear and accurate. For Britain and Europe I've found the *Collins Bird Guide* by Lars Svensson to address the challenges of northern birdsong well. For north America, *The Sibley Guide to Birds* actually omits plumage details in preference to descriptions of voice. Richard Grimmett, Carol and Tim Inskipp's guides to the birds of India are comprehensive, while Pratt and Beehler is the bible for Papua New Guinea.

Chapter 3: Nature Tells us Stories

14. Note that a home range (two words) can sometimes refer to the total area inhabited by a species or subgroup, and is therefore far larger than the homerange (one word) of a particular individual, pair or family group.

15. Eugene S. Morton. (1986). 'Predictions from the Ranging Hypothesis for the Evolution of Long Distance Signals in Birds'. *Behaviour*, Vol. 99, No. 1/2, October, pp. 65–86.
    Marc Naguib. (1996). 'Auditory distance estimation in song birds: Implications, methodologies and perspectives'. *Behavioural Processes*, Vol. 38, Issue 2, November, pp. 163–168.

16. Gisela Kaplan. (2019). *Australian Magpie: Biology and Behaviour of an Unusual Songbird*. CSIRO Publishing, ISBN: 9781486307241.

Chapter 4: Hearing Sentience

17. Carl Safina. (2016). *Beyond Words, What Animals Think and Feel*. Souvenir Press, ISBN: 9780285643475. In this Ted talk, Carl describes empathy with animals around 7:30. Aw, heck, watch the whole thing, it is so worth it. ⁗

18. Hollis Taylor. (2017). *Absolute Bird*. Double CD and 44 page book. ⁗

19. Hollis Taylor. (2017). *Is Birdsong Music? Outback Encounters with an Australian Songbird*. Indiana University Press, ISBN: 025302666.

20. View sequencing analysis. ⁗

21. View spectrograms. ⁗

22. BBC segment on the speed of dragonfly perception. ⁗

Chapter 5: Voices of the Land

23. Richard A. Noske. (1998). *Social Organisation and Nesting Biology of the Cooperatively-breeding Varied Sittella Daphoenositta chrysoptera in North-eastern New South Wales.*

24. Tim Low. (2014). *Where Song Began: Australia's Birds and How they Changed the World.* Penguin Random House. ISBN: 9780143572817.

25. John L. Read. (2011). *The Last Wild Island: Saving Tetepare.* Page Digital Publishing Group, ISBN: 9780980760002.

Chapter 6: Listening to Deep Time

26. The Greenish Warbler is a well-studied example. &

27. Mark McDonald, Sarah Mesnick and John Hildebrand. (2006). *Biogeographic characterization of blue whale song worldwide: Using song to identify populations.*

28. M. G. Ritchie and J. M. Gleason. (1995). *Rapid evolution of courtship song pattern in Drosophila willistoni sibling species.*

29. Kate Lynes, Leo Joseph and J. Scott Keogh. (2009). *Multi-locus phylogeny clarifies the systematics of the Australo-Papuan robins (Family Petroicidae, Passeriformes).*

30. View sonogram analysis. &

31. Giuseppe Boncoraglio and Nicola Saino. (2006). 'Habitat structure and the evolution of bird song: a meta-analysis of the evidence for the acoustic adaptation hypothesis.' *Functional Ecology*, Vol. 21, No. 1 (Feb., 2007), pp. 134–142.

Chapter 7: Sonic Strategies

32. The Australian anthropologist W. E. H. Stanner coined the term 'everywhen' to highlight that 'The Dreaming' of Aboriginal thought could not be defined in time; "it was, and is, everywhere".

33. Tim Low, *Where Song Began*, Chapter 3: The First Song.

34. The raptor mimicry of lyrebirds and bowerbirds has a parallel in the northern hemisphere Jays. European Jays, plus north American Blue and Steller's Jays, are recognised for skilful mimicry of local falcons, hawks and buzzards, with their vocalisations being associated with intruder mobbing. These jays are part of the large corvid family, and not closely related to the lyrebirds and bowerbirds, hence their strategy of defensive mimicry has likely evolved independently.

35. Tim Low, *Where Song Began*, pp. 44–49.

36. Songbird species number among the most populous in the world. &

37. View graph. &

38. Tobias Riede, Chad M. Eliason, Edward H. Miller, Franz Goller, and Julia A. Clarke. (2006). 'Coos, booms, and hoots: The evolution of

closed-mouth vocal behavior in birds.' *Evolution*, June 2016 DOI: 10.1111/evo.12988.

39. Penny van Oosterzee. (1997). *Where Worlds Collide: The Wallace Line.* Reed Books Australia, ISBN: 0730114702.

Chapter 8: The Mind of Nature

40. Almo Farina and Stuart H. Gage (editors). (2017). *Ecoacoustics: The Ecological Role of Sounds.* Wiley, ISBN: 9781119230694.

41. While Ecoacoustics and Acoustic Ecology have been written of as synonymous, Acoustic Ecology is a broader discipline that predates the emergence of Ecoacoustics. The World Forum for Acoustic Ecology was established in 1993. (See also Murray Schafer ref in Chapter 2.)

Chapter 9: Avian Co-operation – Birdwaves

42. S. Harsha, K. Satischandra, Enoka Kudavidanage and Sarath W. Kotagama. (2007). 'The benefits of joining mixed-species flocks for Greater Racket-tailed Drongo Dicrurus paradiseus.' *Forktail 23* (2007): pp. 145–148.

Chapter 10: Avian Diplomacy – The Dawn Chorus

43. These delineations of twilight also apply, in reverse of course, to sunset and dusk.

44. Donald Kroodsma. (2005). *The Singing Life of Birds.* Houghton Mifflin. ISBN: 0618405682. Mockingbirds, pp. 68–79.

45. Jennifer Ackerman. (2016). *The Genius of Birds.* United Kingdom: Little, Brown Book Group, ISBN: 9781472114372. USA: Penguin, ISBN: 9781594205217.
Jennifer Ackerman. (2020). *The Bird Way, A New Look at How Birds Talk, Work, Play, Parent, and Think.* United States: Penguin, ISBN: 9780735223028.

Chapter 11: The Listening Peoples

46. I can't locate a video of the live seminar I recall, however, here Mary Graham discusses the foundations of Aboriginal thinking, including the idea (around 7 min) that the character of the land had come into the peoples that lived there. 🔗

47. I've no doubt the idea I'm suggesting in this chapter is one thread in a complex story of how northern societies evolved. A compatible thesis is that presented by Lynne Kelly, which is supported by ample archaeological evidence and equally affirms the significance of sound and listening. As hunter gatherer peoples adopted more settled and agricultural ways of life, they transitioned from societies based

in knowledge, to ones founded on wealth (the control of land and resources) and violence. Knowledge, previously derived from nature and conveyed orally from memory, instead became transmitted through writing. Power, divorced from the natural authority of elders who had acquired it through lived experience, became invested in ruling elites claiming authority from abstract gods, and backed by warrior casts and literary elites. Many Indigenous and Aboriginal societies the world over have never lost their oral traditions and natural respect for elders and wisdom keepers. They retain their connection to knowledge derived from nature and continue to be Listening Peoples.

Chapter 12: The Communicating Biosphere

48. Phil Senter. (2008). 'Voices of the Past: A Review of Paleozoic and Mesozoic Animal Sounds.' *Historical Biology*, Vol. 20, No. 4, December 2008, 255–287.

49. Meta Virant-Doberlet and Andrej Cokl. (2004). 'Vibrational communication in insects.' *Neotropical Entomology* 33(2).

50. James P. Carse. (1987). *Finite and Infinite Games.* New York: Ballantine Books. ISBN: 9780345341846.

51. Joseph Jordania. (2017). *A New Model of Human Evolution: How Predators Shaped Human Morphology and Behaviour.*

52. Kelly L. Davis and Jennifer A. Clarke. (2019). 'A Tasmanian devil call encodes identity and decreases agonistic behaviour.' *Bioacoustics* 29(5): 1–16.

53. Manuel Molles. (2002). *Ecology, Concepts and Applications*, pp. 360–361. McGraw Hill, ISBN: 007029416X.

54. Aniruddh D. Patel. (2010). *Music, Language and the Brain.* Oxford University Press, ISBN: 9780199755301.

55. Aniruddh D. Patel. (2017). 'Why Doesn't a Songbird (the European Starling) Use Pitch to Recognize Tone Sequences? The Informational Independence Hypothesis.' *Comparative Cognitive & Behaviour Reviews*, Vol. 12, 19–32.

56. S. A. Walsh, P. M. Barrett, A. C. Milner, G. Manley and L. M. Witmer (2009). 'Inner ear anatomy is a proxy for deducing auditory capability and behaviour in reptiles and birds.' *Proceedings of the Royal Society B: Biological Sciences*, 276 (1660): 1355–1360.

57. The fact of the gathering wave of biodiversity loss in our current time is a sure sign that human impacts on our planet constitute such a cataclysmic event.

58. An (admittedly visual) example from the lyrics of Canadian singer-songwriter Bruce Cockburn (3rd verse). 🔗

Chapter 13: An Ecological Future

59. Jane Goodall, with Phillip Berman. (1998). *Reason for Hope: A Spiritual Journey.* Grand Central Publishing, ISBN: 9780446676137.

60. The Seville Statement on Violence. &

61. The Planetary Boundaries concept is a way of comprehending what is required for humankind to live sustainably on Earth. Developed by 28 internationally renowned scientists, this ground-breaking research identified for the first time, the nine processes that regulate the stability and resilience of the Earth system. &

62. Sturt W. Manning, Cindy Kocik, Brita Lorentzen and Jed P. Sparks. (2023). *Severe multi-year drought coincident with Hittite collapse around 1198–1196 BC.* &
Kyle Harper. (2018). *The Fate of Rome: Climate, Disease, and the End of an Empire.* Princeton University Press, ISBN: 9780691166834.
Mark Fiege. (2013). *The Republic of Nature: An Environmental History of the United States.* Weyerhaeuser Environmental Books, ISBN: 9780295993294.
Jared Diamond. (1997). *Guns, Germs and Steel.* W. W. Norton, ISBN: 0393317552.

63. A supplementary chapter to this book dicusses the sonic strategies that may have shaped Hominin evolution. &

64. Mark Fisher attributes a similar quotation "it is easier to imagine an end to the world than an end to capitalism", to both Fredric Jameson and Slavoj Zizek.

65. Edward O. Wilson. (2006). *Half Earth: Our Planet's Fight for Life.* Liveright / W. W. Norton, ISBN: 9781631490828.

66. Thomas Berry. (1999). *The Great Work: Our Way into the Future.* Three Rivers Press. ISBN: 9780609804995. The Earth family is a phrase I believe he coined.

Acknowledgements

67. Sarah's ceramic work on instagram: @sarahkoschak_ceramics. &

68. The Australian Wildlife Sound Recording Group. &

69. Geoff Sample's Wildsong. &

70. Lang Elliott, Music of Nature. &

71. TEDx talk video. &

72. The Australia Earth Laws Alliance is part of a global movement to rethink our place within nature, using the law not simply to regulate, but to embody our custodial relationship with all life. &

Index

Audio resources are listed in **bold**.

www.ingramcontent.com/pod-product-compliance
Lightning Source LLC
Chambersburg PA
CBHW032119020426
42334CB00016B/1007